THE

ENERGY
READER

OVERDEVELOPMENT AND THE DELUSION OF ENDLESS GROWTH

EDITED BY

TOM BUTLER, DANIEL LERCH, AND GEORGE WUERTHNER

INTRODUCTION BY RICHARD HEINBERG

WATERSHED MEDIA

CONTEMPORARY ISSUES SERIES

Published by the Foundation for Deep Ecology in collaboration
with Watershed Media and Post Carbon Institute.

Foundation for Deep Ecology
Building 1062, Fort Cronkhite
Sausalito, CA 94965
www.deepecology.org

Watershed Media
www.watershedmedia.org

Post Carbon Institute
www.postcarbon.org

ISBN 978-0970950093
Library of Congress Control Number: 2012939028

Printed in Canada on recycled paper (100% post-consumer waste) certified by the Forest Stewardship
Council.

Grateful acknowledgment is made to the following authors and publishers that authorized
copyrighted works to be reprinted or adapted for use in *The Energy Reader.*

"Five Carbon Pools" is adapted from *Consulting the Genius of the Place: An Ecological Approach to a New Agriculture*;
© 2011 by Wes Jackson, used by permission of the author and Counterpoint. "Faustian Economics" was
published originally in *Harper's* and later appeared in Wendell Berry's book *What Matters?: Economics for a Renewed
Commonwealth*; © 2010 by Wendell Berry, used by permission of the author and Counterpoint. "Our Global
Ponzi Scheme" was adapted from *Plan B 4.0: Mobilizing to Save Civilization*; © 2009 by Lester R. Brown, used
by permission of the author and W.W. Norton & Company. "Alternative Energy Challenges" was adapted from
David Fridley's essay "Nine Challenges of Alternative Energy" in *The Post Carbon Reader*; © 2010 by David
Fridley, used by permission of the author and Watershed Media. "Malevolent and Malignant Threats" by R. James
Woolsey was adapted from the author's chapter in *Climatic Cataclysm: The Foreign Policy and National Security
Implications of Climate Change*; © 2008 by the Brookings Institution, used by permission of the author and the
Brookings Institution. "Progress vs. Apocalypse" by John Michael Greer was adapted from *The Long Descent: A
User's Guide to the End of the Industrial Age*; © 2008 by John Michael Greer, used by permission of the author and
New Society Publishers. Portions of Richard Bell's "Nuclear Power and the Earth" are adapted from his book,
Nukespeak: The Selling of Nuclear Technology from the Manhattan Project to Fukushima (Sierra Club Books, 2011); ©
2011 by Richard Bell, used by permission of the author. "The Whole Fracking Enchilada" appeared originally
in the September/October 2010 issue of *Orion* magazine, and was published subsequently in *The Best American
Science and Nature Writing, 2010* (Houghton Mifflin); © 2010 by Sandra Steingraber, used by permission of the
author. "Re-tooling the Planet" was adapted from the report *Retooling the Planet* (lead authors Diana Bronson, Pat
Mooney, and Kathy Jo Wetter) that ETC Group prepared for the Swedish Society for Nature Conservation, used
by permission of ETC Group. "Tar Sands, Pipelines, and the Threat to First Nations" by Winona LaDuke with
Martin Curry was published originally in *Indian Country Today*; © 2012 by Winona LaDuke, used by permission
of the author. "Sweet and Sour: The Curse of Oil in the Niger Delta" by Michael Watts is adapted from *Curse of
the Black Gold: 50 Years of Oil in The Niger Delta* (powerHouse Books, 2008); © 2012 by Michael Watts, used by
permission of the author. "Outsourcing Pollution and Energy Intensive Production" by Vandana Shiva includes
material that has appeared elsewhere; used by permission of the author. "Three Steps to Establish a Politics of
Global Warming" was adapted from an essay of the same name that appeared originally on TomDispatch.com; ©
2012 by Bill McKibben, used by permission of the author. Amory Lovins adapted his essay "Reinventing Fire"
from *Reinventing Fire: Bold Business Solutions for the New Energy Era* (Chelsea Green Publishing), produced by
Lovins and his Rocky Mountain Institute colleagues; © 2011 by Rocky Mountain Institute.

DISCLAIMER

Energy policy is contentious. Thoughtful, well-meaning individuals can and do disagree, sometimes vigorously. The editors of *Energy* and *The Energy Reader* believed that it was crucial to present a variety of voices with disparate perspectives. The inclusion of an essay or photographs in this volume by a particular author or photographer does not constitute his or her endorsement of other contributors' views or the educational objectives of the Foundation for Deep Ecology. Similarly, the inclusion of an author's or photographer's work does not constitute endorsement of his or her policy positions or political agendas on the part of the Foundation for Deep Ecology or its publishing partners.

To recover from our disease of limitlessness, we will have to give up the idea that we have a right to be godlike animals, that we are potentially omniscient and omnipotent, ready to discover "the secret of the universe." We will have to start over, with a different and much older premise: the naturalness and, for creatures of limited intelligence, the necessity, of limits. We must learn again to ask how we can make the most of what we are, what we have, what we have been given.

—WENDELL BERRY

DEDICATION

For the wild creatures whose habitat is being destroyed by a rapacious energy economy, and for the children whose breathing is labored due to pollution from fossil fuels. May a future energy economy that mirrors nature's elegance arrive soon enough to relieve their suffering.

CONTENTS

FOREWORD

DOUGLAS R. TOMPKINS

This book is the companion reader to *Energy: Overdevelopment and the Delusion of Endless Growth*. *Energy* was several years in the making and builds on an earlier work, *Plundering Appalachia: The Tragedy of Mountaintop-Removal Coal Mining*, which our foundation published in 2008. For nearly twenty years the Foundation for Deep Ecology has produced a series of hard-hitting books on environmental issues as a means of stimulating activism and helping build the effectiveness of the conservation movement.

Our publishing efforts started with the book *Clearcut*, in which we illuminated how industrial forestry practices were, essentially, "mining" forests. Through the years we have continued scrutinizing the "mining" industry—whether it targets soils, trees, coal, or, as witnessed in this volume, the whole range of energy sources including fossil fuels, wild rivers, biomass, and so on.

Energy started out as a book on the tar sands oil industry in Alberta. Following initial research trips to Canada, a review of the energy literature, and communication with activists opposing the cancerous expansion of tar sands development, it became apparent that our editorial focus should expand. The quest for energy of all kinds is laying waste to ecosystem after ecosystem, landscape after landscape. Thus we needed a comprehensive overview of energy extraction and generation—from hydroelectric, to nuclear, to fossil fuels, to biomass, to solar— to consider how various forms of energy development affect biodiversity. None are benign. All have their impacts; all reduce beauty, ecological integrity, and environmental health to one degree or another.

Seen and unseen, energy-related impacts are ubiquitous. Even the chemical composition of the atmosphere records the way that humanity is using energy. As soon as one looks beneath the surface—behind the light switch or gasoline pump—one sees an energy economy that is toxic to nature and people. Everywhere the ways we generate, use, and waste energy are the driving force behind humanity's unconscious but persistent and seemingly implacable assault on wild nature. Energy holds up the entire "scaffolding of civilization," to borrow Wes Jackson's phrase, and in turn the scaffolding of civilization is necessary to support the entire energy economy, from the old-fashioned oil pump-jack to the shiniest new thin-film photovoltaic array.

A deeper look at the energy picture reveals ugliness. The techno-industrial growth culture's relentless assault on beauty has real and growing costs, even if they are sometimes difficult to account for in monetary terms. The simple fact is that a landscape disfigured by a coal mine or drill pads or giant industrial windmills or 200-foot-high electric transmission towers is a landscape where ecological and aesthetic values have been diminished—representing a discount on our well-being that can never be fully offset by immediate economic benefits. The beauty of wild nature has no price and therefore is not entered into the formulas of economists obsessed with quantitative measures of value.

Necessarily, the photos in *Energy* depict damaged landscapes and suggest the related biodiversity loss, degraded soils, corporation-influenced politics, and polluted communities that spring from the current, rapacious energy economy. No one enjoys looking at the images of such devastation, but look at them we must if we are to understand the magnitude of the current problem, and begin formulating policies that are not merely Band-Aid reforms to the present system.

What spurs our own energies to produce these kinds of books is our belief that the foundational worldview and epistemology of industrial society must be examined deeply. To confront a problem effectively one must first develop an adequate systemic analysis and critique. By working toward widespread "energy literacy" we can help build what we call the intellectual infrastructure necessary for society to conduct such an analysis. Upon that foundation we can craft the strategies and programs—implemented incrementally but oriented toward fundamental restructuring—to create an appropriately scaled energy economy that is attuned to particular places but generally mirrors nature's own economy, which produces beauty and diversity and no waste.

In any conversation about the future direction of energy policy, it is crucial to face the hard reality that the loss of biodiversity and the unfolding global extinction crisis are the direct results of the overdevelopment of civilization, including the human demographic explosion. Our insistence upon continually growing the economy has left the natural world, upon which civilization is wholly dependent, in a precarious and degraded condition. The ugly manifestations of planetary overshoot are to be seen across the globe. Overscaled, outsized, overdeveloped are the adjectives of the age. If one cannot see it, then one is not looking.

Today there is growing recognition that wild nature is in a survival struggle, and that the fate of civilization is bound to the fate of the oceans and the climate. It appears that we are muddling toward ecological Armageddon and yet root causes are seldom explored and discussed.

The current energy economy is built not just upon coal mines, nuclear plants, and oil tankers—it is also built on ideas. The "arrogance of humanism," as David Ehrenfeld has helpfully encapsulated the problem, is at the heart of the eco-social crisis. The notions that people are in charge of nature, can effectively manage the Earth solely for human ends, and can escape the ecological consequences of their own actions are intellectually indefensible. Yet the entire collective human enterprise continues to be steered down these cognitive dead ends. Climate change, the extinction crisis, the depletion of resources of all kinds, and resulting economic social crises can be seen as the inevitable products of our collective delusional thinking.

People may be tired of doomsday predictions—but it is irresponsible, in our view, to turn a blind eye to the evidence before us. The visual testimony in *Energy* is, we believe, compelling. So for those who are willing to see this evidence and to be moved by it, the first important step is to understand the root causes of the crisis which ensnares us all. That means, in our view, that we must push ourselves along the learning curve of energy literacy—and invite every thoughtful citizen to join us—to understand the real causes of the impacts of our energy policies and programs.

In addition, it is essential that we shine a bright light on the delusional dogma of unending growth if we are to remake our economies under different terms, and share the planet in a way that allows evolution to flourish again. Anything short of that will lead us to darker and darker days.

In each of our publishing projects we work with the leading thinkers, activists, and NGOs concerned with the book's topic. *The Energy Reader* is no exception, and we anticipate that the distinguished group of scholars and activists here will help readers quickly gain a solid understanding of energy fundamentals. It is not a book about "solutions." Rather, its aim is to place the energy question in the context of the larger predicament facing humans and nature.

We have been pleased to collaborate on this effort with the staff and various fellows of the Post Carbon Institute. The experts and activists who contributed to this book may disagree, and frequently do. The diversity of voices and opinions presented is representative of multiple streams of thinking within the energy debate. But it is interesting to note that leading thinkers on energy from across the spectrum tend to agree that conservation should be the central organizing principle of energy policy.

One may believe the energy future will be shaped by geological constraints and scarcity or that innovation and technology will enable growth to continue for some time. In either case, there is every reason to embrace energy conservation and a reduction of humanity's ecological footprint as the wise and ethically responsible course of action.

Ultimately, this volume is an invitation to activism for everyone who hopes for a future in which humanity gains the humility to be a plain member and citizen of the ecosphere, rather than presuming to be Manager of the Earth. If you care about the human future and about other species whose very existence is at stake, then join the rest of us who are actively working to preserve beauty, wildness, the health of nature, and human well-being. Overhauling a toxic energy economy and rebuilding it with one that supports the values of beauty and biodiversity is vital— and meaningful—work in which every citizen can play a part.

DOUGLAS TOMPKINS *is a longtime wilderness advocate, mountaineer, organic farmer, and activist. He founded The North Face outdoor gear retailer and cofounded Esprit clothing. Since retiring from business and moving to Chile in 1990, he has worked to create large-scale protected areas in Chile and Argentina. He and his wife, Kristine Tompkins, have also restored several degraded farms. Through a family foundation, Doug Tompkins has been a supporter of activist groups in North and South America, and he has helped produce numerous campaign-related books on topics such as industrial forestry, motorized recreation, corporate agriculture, and coal mining.*

INTRODUCTION

RICHARD HEINBERG

E nergy is at the core of the human predicament in the twenty-first century. Extracting fossil fuels poisons landscapes, fragments habitat, and destroys beauty. Burning those fuels is changing the chemical composition of the global atmosphere and accelerating climate change. At the same time, spiraling fossil fuel prices—resulting from depletion of the highest-grade and most easily accessed hydrocarbon resources—have contributed to a worldwide financial crisis that threatens global stability. Not only are transport costs rising, threatening globalized supply chains, but soaring energy prices also drive up food prices, leading to increasing social unrest around the world.

As conventional oil and gas deplete, energy companies are forced to spend more and more to search for and produce resources that are farther afield, that are more technically challenging to access, and that pose serious risks to ecosystems. In their increasingly desperate search for "extreme energy," oil and gas companies must operate at the margin of their technical capabilities. Under these circumstances, accidents are not only more likely to happen, but are often far more disastrous when they do: Recall the Deepwater Horizon catastrophe in the Gulf of Mexico in 2010, and imagine a similar or larger accident happening hundreds of miles off the coast of Alaska in rough arctic seas. Indeed, the entire project of globalized industrial civilization—which took root and dramatically expanded during the twentieth century as cheap energy drove production, trade, and population growth—now seems imperiled as energy and ecological limits come into view.

It's tempting to take the micro-view and look for ways to target each of our energy problems with a technical fix. Can't we improve the energy efficiency of vehicles, insulate our buildings, and develop renewable energy sources? Yes, of course. Can't we regulate the fossil fuel industry better, and allow the vast, recently unlocked North American reserves of shale gas and shale oil to be produced responsibly? Possibly. We could do all of those things, and many more besides, to lessen the current energy economy's impacts on natural and human communities—and still there would remain serious obstacles ahead.

Why? Let's zoom out from the details of our dilemma and take in the big picture. As we do, two fundamental problems become clear:

FIRST: We have exceeded global levels of energy consumption that are sustainable. The sheer scale of our energy use today is fantastic when compared with that in any era of history. Today the human population is roughly seven times larger than it was just prior to the Industrial Revolution—a dramatic and dangerous population growth trajectory—but we use 30 times as much energy. On average, each human today uses the energy equivalent of 360 gallons of gasoline annually, which translates to about as much energy as a typical person would expend in ten years of hard labor. On average, each American commands the services of roughly 150 "energy slaves." (In other words, if all power consumed by each of us had to be supplied by human muscles, it would take 150 people working 24/7 to supply it).

And still we want more! We have come to depend on economic growth in order to ensure more jobs and higher profits each year, and economic growth implies increased use of energy. Yet producing incremental additions to global energy supplies has become a task of monumental proportions, requiring soaring amounts of investment capital as well as swelling streams of raw materials (not only coal, oil, and gas, but also water, steel, copper, uranium, neodymium, lithium, gallium, aluminum, etc.) as well as burgeoning armies of trained personnel. In short, we have gotten used to an economy that is overpowered and that demands still more power each and every year.

SECOND: We have created an energy infrastructure that has overpowered natural ecosystems, thereby threatening the future of many species—our own included. The vast scale of our production of highly concentrated fuels enables us to use tools of a colossal nature. Diesel-powered ocean trawlers overwhelm the ability of commercial fish species to rebound. Diesel-powered shovels rip apart mountains to get at the coal and other minerals buried inside. A billion cars and trucks and tens of thousands of jet aircraft and coal- and gas-fired power plants spew carbon into the air, undermining climate stability. Chain saws and bulldozers level 13 million hectares of forest per year, while diesel-powered paving machines turn thousands more hectares of agricultural land, forest, and habitat into highways, runways, and parking lots. The United States alone has paved 6.3 million kilometers (3.9 million miles) of roads, enough to circle the Earth at the equator 157 times.

All this energy consumption adds to the number of humans that Earth can support—in the short term. We grow immense amounts of food through fossil fuel-powered industrial agriculture, and we transport goods around the world by airplane, train, truck, and ship to compensate for local scarcity. But over the longer term, as natural ecosystems decline, we are degrading Earth's ability to support human beings. Exactly when these two trends will converge, and how, is still a matter for speculation—but evidence suggests the intersection may be years away, not decades or centuries.

Currently, our strategy for staving off social and ecological collapse is yet more economic growth. That means still more energy production to fuel more industrialization of agriculture, more transportation, more construction, and more manufacturing. But how do we propose to increase energy supplies? The solutions being put forward by the energy industry and most governments include: applying more advanced technology to the exploitation of marginal fossil fuels (tar sands, deepwater oil, shale oil, shale gas, and others); building more nuclear reactors and developing third- and fourth-generation reactor technologies; and tapping renewable energy sources such as wind, solar, geothermal, tidal, and wave power. Nearly everyone agrees that we should also use energy much more efficiently than we do now. But even factoring in realistic efficiency gains, official agencies such as the International Energy Agency and U.S. Department of Energy predict increasing demand for energy for as far into the future as their forecasts can peer.

This book argues that, while our choices about which energy resources we use are important, each and every option has costs. Even energy efficiency has costs: It is subject to the law of diminishing returns (each further increase in efficiency tends to cost more than the previous one), while energy saved in one part of the economy will tend to be used in another. And the costs of increasing our energy production are, in more and more instances, exceeding the benefits. We have reached a point of crisis with regard to energy, a point where the contradictions inherent in our growth-based energy system are becoming untenable, and where its deferred costs are coming due. The essential problem is not just that we are tapping the wrong energy sources (though we are), or that we are wasteful and inefficient (though we are), but that we are overpowered, and we are overpowering nature.

These conclusions have no constituency among the powerful. Politicians and business leaders seem interested only in finding ways to increase energy production and consumption. But if what we are saying in this book is true, the only reasonable path forward is to find ways to use less energy.

For the already-industrialized world, energy consumption is at such high levels that substantial reductions would still leave plenty of room for the enjoyment of modern conveniences. For less-industrialized countries, where hundreds of millions live with very little electricity or liquid fuel, it is essential that "development" be redefined in terms of sufficiency and quality of life, instead of being measured in numbers of cars and highways, and in tons of food, raw materials, and manufactured products exported.

In short, our task in the twenty-first century is to scale back the human enterprise until it can be supported with levels of power that can be sustainably supplied, and until it no longer overwhelms natural ecosystems. Undoubtedly, that enterprise will in the end consist of fewer people using less, on a per capita basis, than is currently the case. As we power down, we will find ways to use the technologies and scientific understandings developed during our brief, unsustainable, and probably unrepeatable period of high energy use in order to make the inevitable energy decline survivable and perhaps even salutary. But power down we must.

This book is the companion reader to *Energy: Overdevelopment and the Delusion of Endless Growth*, a photographic tour of the world of energy illustrating the costs and trade-offs of constantly expanding efforts to fuel industrial processes. Pictures often tell a story in ways that words cannot, and we the editors felt that the paper and printer's ink required for that volume were not only justified but required. It is one thing to describe verbally the results of tar sands mining in Alberta. It's quite another to see the shocking images of Canada's boreal forest turned to a blasted wasteland by brontosaurus-sized machines clawing ever deeper into a constantly expanding environmental sacrifice zone.

Energy (and *The Energy Reader*) features essays by a diverse collection of authors who do not necessarily agree with one another on every point. Their contributions were selected to highlight a range of issues related to energy production and consumption, and especially the environmental consequences of energy use, in the twenty-first century. The book is organized to illuminate topics according to a sequential and cumulative logic. However, readers may feel free to dip in anywhere, as their interest leads them. Each essay makes its own case on its own terms.

What emerges from this verbal and pictorial mosaic is an impression of a planet and a society in crisis—a crisis of overconsumption on one hand, and of overwhelming impacts on increasingly fragile natural systems on the other. Our goal is to help change the national and global conversation about energy—to help it evolve quickly from one of how to grow energy production to one of how to shrink our appetites to fit nature's ability to sustain itself.

Take a good look at what it takes to power our human world of cities and machines. Think about the tens of millions of years' worth of fossil fuels we are burning in mere decades; about the billions of tons of geologically stored carbon we are releasing into the atmosphere; about the landscapes we ravage, the water we foul, the air we pollute, and the species we drive into extinction in order to fuel our industrial mega-machine.

Then ask yourself: Is all of this really necessary? Couldn't we just use less?

Part One

A DEEPER LOOK AT
THE ENERGY PICTURE

INTRODUCTION

Energy Literacy

Just as a person doesn't need to be a doctor to recognize sickness in a family member, one doesn't need to be an expert to see the most blatant negative effects of the current energy system. Many of the photos in *Energy* starkly depict the garish wounds that industrial growth society, through its incessant demand for more power, is inflicting on the natural world. Anyone who takes a moment to look can see these threats to wildlife habitat, free-flowing rivers, air and water quality, and human welfare.

But some consequences of the current energy economy are subtle and require a closer look to discern. Even less visible is the range of ideas and assumptions—the worldview—behind our energy choices. These choices collectively shape the society that we live in and the range of options available to future generations. Will our energy economy foster beauty, promote social equity, and leave enough room for wild nature to flourish? Or will it produce more ugliness, political corruption, habitat destruction, and commodification of nature? Will it unleash climate chaos, trigger mass extinction of our fellow creatures, and make human habitation untenable across large swaths of the globe?

These are the big questions that undergird any discussion of energy policy. People living in the overdeveloped world have become accustomed to consistent and affordable power; few citizens give much thought to energy matters until there is a supply interruption or price hike. With human population increasing rapidly and the "easy" fossil energy resources already exploited, per capita energy availability is almost certain to decrease in the coming decades. More landscapes and seascapes will be disfigured in the mad rush to continue powering economic growth. Costs will rise, and various parts of the energy system may become unstable.

Today, energy policy is everybody's business, and everyone should understand the policy choices and trade-offs under discussion. Which energy sources cause the most damage to natural habitats and which produce the most greenhouse gas pollution? What are the health and economic ramifications of mountaintop-removal coal mining? Will new technology and renewables solve all our energy problems? Can efficiency keep pace with consumption in a growth society?

Energy literacy is critical if citizens are to think substantively about these and other energy policy questions. It is necessary both to understand the key concepts that inform energy debates and to be familiar with the current terrain of the energy landscape. This entails some study on the part of individuals, but it is crucial if society is to move beyond sloganeering ("drill, baby, drill") in our public discourse.

A solid grounding in energy fundamentals is the first step toward thinking more deeply about root causes and systemic problems in the current energy economy. For the health of wild nature and human communities, there is no task more urgent than promoting widespread energy literacy so that we can begin charting a course that moves civilization toward a durable, nature-friendly energy future.

A DEEPER LOOK AT
THE ENERGY PICTURE

ENERGY AND THE SCAFFOLDING OF CIVILIZATION

Virtually every aspect of modern society—manufacturing, transportation, construction, heating, lighting, communication, computing, farming, preparing for and waging war—depends on a continuous supply of abundant, inexpensive energy. In essence, energy supports the entire scaffolding of civilization.

Societies with complex social organization have existed for thousands of years, but until the discovery and widespread exploitation of fossil fuels, human economies were powered by the productive capacity of local ecosystems, augmented by regional trading networks. Today, a hyper-complex web of mining, manufacturing, distribution, and marketing systems is the foundation of commercial enterprise. The everyday products of mass consumption in affluent countries would disappear without these interdependent systems, all of which require energy.

A development model that demands economic growth depends upon ever-increasing energy consumption. Since the Industrial Revolution, exploitation of fossil fuels—the onetime windfall of ancient biological capital processed by geological forces—has precipitated exponential population and economic growth. But if the scaffolding of civilization is constructed on a flawed foundation—the idea that perpetual growth is possible on a finite planet—then it cannot remain standing indefinitely.

WHAT IS ENERGY?

Though we cannot hold a jar of pure energy in our hands or describe its shape or color, it is nevertheless the basis of everything. Without energy, nothing could happen; matter itself could not exist in any meaningful sense. But because energy as such is so elusive, physicists and engineers define it not in terms of what it *is*, but what it *does*—as "the ability to do work," or "the capacity to move or change matter."

In traditional societies, most useful energy came from the sunlight annually captured by food crops and forests; people exerted energy through muscle power and obtained heat from firewood. Modern industrial societies obtain enormously greater amounts of energy from fossil fuels, nuclear power, and hydroelectric dams, and they exert energy through a vast array of machinery. Industrial energy production is essential to every aspect of modern life, but no matter how far our technology for capturing or using energy advances, energy itself always remains the same.

In the nineteenth century, physicists formulated two fundamental laws of energy that appear to be true for all times and places. These are known as the First and Second Laws of Thermodynamics. The First Law is known as the law of *conservation*. It states that energy cannot be created or destroyed, only transformed. Think of energy as a singular reality that manifests itself in various forms—nuclear, mechanical, chemical, thermal, electromagnetic, and gravitational—and that can be converted from one form to another.

The Second Law states that in every energy conversion, some energy is dissipated (typically as heat). When the gas gauge in a car moves from "full" to "empty," it may appear that the energy that is chemically stored in gasoline is being *consumed*. But all the energy that was originally present in the gasoline still exists. In reality, the stored energy is merely being released and doing some work as it moves from a condition of higher concentration to one of lower concentration. It is converted from chemical storage (via the atomic electromagnetic bonds within hydrocarbon molecules) to mechanical motion and heat (as combustion within the engine's cylinders pushes the car forward and also increases the rate of motion of molecules in the cylinder and the surrounding environment).

We might be able to get some work out of the "wasted" heat being given off by the burning of gasoline in the car engine; but heat tends to radiate quickly into the general environment, so we would have to use that heat both immediately and close to the engine. If we could gather up all the heat and mechanical energy that was released by burning the tankful of gasoline, it could do just as much work for us yet again; but the act of reconcentrating and storing it would require more energy than we could regather. Thus, in effect, available energy is always being lost.

The Second Law is known as the law of *entropy* (entropy is a measure of the amount of energy no longer practically capable of conversion into work). The Second Law tells us that the entropy within an isolated system inevitably increases over time. Energy that is sufficiently concentrated (relative to background energy levels) so that it can do work for us is called a *source*. There are two kinds of energy sources: *flows* (examples include sunlight, winds, and rivers) and *stocks* (a word that in this context refers to energy chemically stored in substances such as wood or fossil fuels). Flows tend to be variable, whereas stocks deplete.

NET ENERGY

A business may have high gross receipts and still go broke; it is the net, the profit after costs are subtracted, that determines viability. For any potential energy resource, the fundamentals are the same. How much energy is available after subtracting the energy costs to extract, process, and deliver the resource? To know how much energy from a particular source can actually be deployed by society, we must factor in both the production costs and the system costs—that is, the energy required to make energy available to the end user. With gasoline, for instance, this calculation would include energy costs related to oil exploration, drilling, refining, transportation, and the infrastructure that supports each step of the process. With coal-derived electricity, the calculation would include the life cycle from mine to power plant to electric grid.

Experts who study this use the terms "net energy ratio" or "energy returned on energy invested" (EROEI). Decades ago when the most accessible reserves were drilled, an oil company might produce 100 barrels of oil or more for each barrel's worth of energy invested. Declining oil field productivity has brought the average net energy ratio for conventional oil down to approximately 20:1 globally, with more remote or hard-to-refine oil significantly worse. For fossil energy generally, the trend is downward despite technological advances in exploration and drilling. For biofuels, the net energy ratio is lower still. Some studies suggest that corn-derived ethanol actually has a *negative* net energy ratio—that is, more energy than a gallon of ethanol can deliver is used to produce a gallon of ethanol. Sugarcane-based ethanol has a superior net energy ratio, but it is still low compared to fossil fuels.

Any produced energy resource can be analyzed for its net energy ratio, although the process raises a difficult question: What are the boundaries of consideration? For example, when tallying the energy required to build a solar photovoltaic panel, what should be included in the accounting? The energy needed to mine the bauxite for the aluminum frame? The energy needed to manufacture the heavy equipment that did the mining? The energy needed to construct the factory that produced the panel? Where the boundaries are drawn affects the final net energy ratios.

A society that depends on inexpensive energy to maintain a high standard of living and constant growth faces a predicament—it cannot maintain itself over the long run without high net energy fuels. Oil, natural gas, and coal have provided a huge, high-quality energy subsidy to the modern world. That subsidy, which has enabled human population and wealth to grow exponentially, is based on finite resources and cannot continue indefinitely. Renewable energy sources, excluding hydropower, are generally more diffuse and have lower net energy ratios than fossil fuels. If high net energy sources are in decline, and no reasonable replacements are available, the result may be a painful restructuring as society rearranges economic activity to fit a diminishing energy supply.

ENERGY DENSITY

Different fuels contain more or less potential energy per unit of weight or volume, and even within fuel types, such as wood or coal, the heat value varies. Anthracite packs more energy than bituminous coal, and putting oak rather than pine in the woodstove before bedtime makes a big difference in how warm the house will feel on a winter morning. The fossil fuel age has been such a bonanza because oil and coal are extremely energy-dense fuels. They have benefited from the long work of geological processes to concentrate the carbon molecules from ancient plant and animal matter.

On average, coal has approximately twice the energy density of wood. Liquid fuels refined from petroleum including gasoline, kerosene, diesel, and heating oil all contain more than three times the energy value of wood. It is no accident that when human societies have had the opportunity to transition from locally harvested biomass to concentrated fossil energy fuels they have chosen to do so.

The miraculous quality of fossil fuel energy density is easy to understand if one imagines trying to push an automobile for twenty miles. Given enough time, and some help from athletic friends, it would be possible to push a 3,000-pound car that distance. But it would require a tremendous amount of effort. And yet a mere gallon of gasoline (which, despite recent price increases, still costs far less in the United States than an equivalent amount of good coffee) can easily power a car that far in the time it takes to drink a mocha latte. The fact that renewable energy is, in general, more diffuse than fossil fuel presents the primary challenge to transitioning from the current energy economy to a renewables-powered future.

EMBODIED ENERGY

Every material artifact—a carrot bought at the grocery store, the cooler where it was displayed, the supermarket building, the car driven there, and the road network it travels—requires a certain amount of energy in its manufacture, maintenance, and eventual disposal. The methods used to analyze the total embodied energy of manufactured objects vary, but in general, studies over the decades have used life-cycle analysis to quantify embodied energy in computers, household appliances, automobiles, and other common products.

The embodied energy in our physical infrastructure—from water mains and buildings to superhighways and airports—is immense, and thus infrastructure is one of the most important areas where energy use (and associated greenhouse gas pollution) could be reduced. In addition to building smaller, or building less, we can also build differently. Wood, for example, has the lowest embodied energy of common building materials; plastic has approximately six times as much embodied energy by weight, glass 16 times as much, steel 24 times as much, and aluminum a whopping 126 times as much embodied energy as wood. Erecting the scaffolding of civilization took a great deal of energy, and maintaining and expanding it takes more all the time. This vast amount of embodied energy, along with psychological and financial investments in the current energy distribution system, is a key obstacle to fundamental changes in that system.

Another useful metaphor that communicates the idea of embodied energy across a product's life cycle is the "energy train." Take for example that ubiquitous artifact of modern civilization, the mobile phone. To its owner, a cell phone is simply a handy gadget that offers convenience and a feeling of connection. But the phone does not exist in isolation—it isn't a single locomotive chugging down the tracks; rather, it pulls a train of cars behind it, all of which have ecological and energetic costs. Those metaphorical railroad cars are filled with packaging to ship the phone; an advertising industry to inculcate desire for it; a retail store to sell it; a communications network that allows it to function; an assembly plant to build it; factories to manufacture plastic cases and computer chips and other components; mines where copper, silver, and rare earth elements are dug from the ground; the transportation infrastructure to move raw materials; and of course the energy system (oil wells, coal mines, power plants, hydroelectric dams, etc.) that support the entire operation. It is a very long train, and every car being pulled along must be in place for even one mobile phone to make its first call.

ENERGY SLAVES

During the vast majority of our species' history, all work was done by human muscles (sometimes the muscles of human beings enslaved by others). After people learned to domesticate wild creatures, beasts of burden such as oxen and horses added to our ability to harness the Sun's energy—captured by plants and channeled into the muscles of work animals. This relationship between domestic animals and the machines we use today is enshrined in the "horsepower" rating of modern engines. More recently, people began using wind and waterpower to amplify human labor. But with the dawn of the fossil fuel age, the average person was able to command amounts of energy previously available only to kings and commanders of armies.

Where people or work animals formerly toiled in the fields, the petroleum-powered machines of industrial agriculture now do the work of growing food. Need to be on the other side of the planet tomorrow? Jet travel can get you there. Want to sit in the sunshine, gamble, and overeat with a few thousand strangers in a gigantic floating hotel? The cruise "industry" can make your dreams come true. Energy-dense fossil fuels make the seemingly impossible or ridiculously extravagant whims of people a reality.

In effect, the modern energy economy provides power equivalent to that of vast numbers of human or animal servants. That is the idea behind the concept of "energy slaves." Although top athletes can do far better, a typical adult male at sustained labor is estimated to produce 75 to 100 watts of power. Calculate the total energy use of an average American and it seems that there are the energetic equivalent of more than 100 energy slaves working around the clock to prop up the easy lifestyle offered by modern civilization.

ENERGY-FUELED POPULATION GROWTH

Humanity's current population explosion is an aberration. During the vast majority of human history, population levels were low and quite stable. Demographer Joel Cohen estimates that from the time our species emerged until roughly twelve thousand years ago, when local agriculture appeared, the population growth rate was less than 1/500th of 1 percent. After the widespread adoption of farming the growth rate ticked up by a factor of ten or more, but for thousands of years thereafter remained at around 1/50th of 1 percent. It took all of human history until the early eighteen hundreds for global population to reach one billion. Then the population doubled—a second billon was added—in just a century or so. Adding the next billion humans to the planet took only thirty years. The next billion, fourteen years. The next, twelve years. After another dozen years, in 1999, world population reached six billion, and the seven billion mark was passed in 2011.

When charted graphically, the human demographic explosion takes the familiar "hockey stick" shape of a classic exponential growth curve. Many factors contributed to demographic expansion, including: the global agricultural revolution in the sixteen hundreds when new foods were shared between continents; the dispersal of scientific and public health knowledge; and increasing urbanization. But central to the runaway population growth of the past two centuries is the incredible windfall of energy that fossil fuels presented to humanity. The ability to command energy, especially highly energy-dense fuels like coal, precipitated the Industrial Revolution and allowed its descendant, the techno-industrial growth culture, to flourish. Food could now be produced in far larger quantities, and local scarcity could be overcome through global transport networks.

Leading ecologists agree that humanity has already surpassed Earth's ecological carrying capacity. Exploiting the onetime reserve of fossil energy has allowed us to temporarily escape the constraints that kept early human population levels in check. Today's global extinction crisis, massive poverty and malnutrition, rising social inequity, and unraveling ecosystems around the globe suggest that the age of abundance is nearly over. As economist Lisi Krall tells her students, "The defining fact of this historical moment is the reality of exponential growth. With exponential growth, if you do the same things as your parents, you'll get entirely different results." Confronting the population problem is the preeminent challenge of our time.

ENERGY-FUELED ECONOMIC GROWTH

World economic activity has historically grown slowly. From the Middle Ages up to the early eighteen hundreds, average per capita income rose only about 50 percent. But since the advent of the Industrial Revolution the pace has picked up, with global per capita income rising more than eightfold in just the last two hundred years.

Energy consumption has also risen dramatically, from under 20 gigajoules (GJ) per person per year in the pre-industrial era to over 75 GJ per person today (and more than 300 GJ per person in the United States). During this period, energy consumption and economic activity have stoked each other in a self-reinforcing feedback loop. Once the fossil fuel tap was opened for the modern world in eighteenth-century Britain, the high-energy content of coal (and, later, oil) enabled unprecedented productivity—spurring more consumption, more demand for energy, and better technology to get at yet more fossil fuels.

Despite the clear link between energy and economic growth, economists have interpreted and normalized growth as resulting from factors such as "market efficiency" and "labor productivity," which (it is assumed) can be counted upon to produce more and more growth, ad infinitum. Policy makers have therefore built dependence on growth into the design of our economic system. Investors demand constant growth and high rates of return. Future growth is assumed to wipe away the debts taken on today by governments, businesses, and households. Most Americans are even betting their retirement savings, sitting in mutual funds on Wall Street, on continued growth.

As the global bonanza of cheap fossil fuels winds down, what will happen to economic growth? Certainly it's possible to get more benefit per joule through smarter use of energy, but using energy efficiency to "decouple" economic growth from energy consumption can only go so far. After the easy efficiencies are found, further efficiency measures often require greater cost for less benefit; and while greater efficiency may reduce costs at first, it can have the effect of spurring yet more consumption.

It's intuitively clear that it takes energy to do things, and modern civilization has exploited high-energy-content fossil fuels to dramatically reshape the living conditions and experiences of billions of people. (Altering the climate and destroying natural ecosystems around the globe were unintended consequences.) In the future, humanity will need to cope with both more expensive energy and less energy available per capita. Maintaining an acceptable level of productivity—let alone growth— may constitute one of society's foremost social, political, technical, and economic challenges.

PEAK OIL AND RESOURCE DEPLETION

Every individual gas or oil well, every oil field, and every oil-producing country experiences a similar lifecycle. After a well is drilled, extraction ramps up to its maximum sustained output and eventually begins to decline as the reservoir is depleted. Then we search for the next well, which is generally a little harder to find, a little more expensive to produce. The price of any fossil energy determines what reserves are economically recoverable, and technological innovations can temporarily reverse the decline or extend well life. But as with any finite, nonrenewable resource—coal, natural gas, uranium, etc.—depletion is inevitable at some point.

In recent years, a large body of literature has begun exploring the many ramifications of "peak oil"—the moment when aggregate global oil production reaches its apex. The late American geologist M. King Hubbert predicted in the mid-1950s that U.S. oil production would reach the top of its production curve around 1970 and then begin to decline. That assessment was remarkably prescient: America's production of crude did peak in 1970 and has been generally declining since, despite the addition of new sources on the Alaska North Slope and in the Gulf of Mexico. The United States, the first great power of the oil age, was also the first nation to explore, exploit, and begin to deplete its conventional oil reserves.

Oil of course is a global commodity. From a global perspective, reaching Hubbert's peak means that roughly half of the world's total oil resources are still in the ground, waiting to be tapped. Practically, however, the second half of the global oil resource is more difficult to access, making it less profitable (in terms of net energy) and more environmentally destructive than the earlier-exploited reserves.

The exact timing of the global oil production peak will be recognizable only in hindsight. Some energy experts predict that the peak will occur sometime during the first two decades of the twenty-first century. Others project continued growth in oil extraction through 2050. Based on data published by the International Energy Agency, global conventional oil production has been essentially flat since 2004, despite record-high prices, and likely peaked in 2006. Increased production of unconventional oil (deepwater oil, tar sands, oil shale, and shale oil) is officially projected to help meet growth in demand in the near future, but some energy experts insist that new production from these sources will be unable to make up for accelerating declines in production from conventional oil fields. Whether peak oil has occurred, is imminent, or remains years or decades off makes little difference to the salient fact: The era of abundant, inexpensive oil is closing, and all the systems for modern life designed around that earlier reality are bound to be affected.

ENERGY SPRAWL

The foremost criterion by which to judge any existing or potential energy source is its systemic ecological impact. A key subset of this analysis is its physical footprint. The useful term "energy sprawl" refers to the ever-increasing area—on land and offshore—that is devoted to energy production. Quantifying the area affected by different energy sources raises challenging methodological questions. It's obvious, for instance, to take into account the drilling pad when considering the energy sprawl impact of oil and gas development. But one should also include the land affected by pipelines, access roads, refining facilities, and other related infrastructure in the calculation. Nuclear power plants occupy a small area relative to their electrical generation output, the smallest physical footprint of any major energy source. That energy sprawl impact grows considerably, however, when one factors in uranium prospecting, mining, processing, nuclear waste disposal, and any new power lines needed for an expanded nuclear industry. Moreover, as past accidents have demonstrated, when nuclear power plants fail, a large area can be contaminated.

Because of their high energy densities, coal, oil, and natural gas have a medium-size footprint if judged on an energy-output-per-acre ratio; but in practice these extractive industries affect a huge and growing area because they dominate energy production, and because of the enormous quantities of energy being consumed. Oil shale development in the American West is a potential area of fossil fuel exploitation that would create massive energy sprawl. Renewables, which harness the diffuse energy sources of wind and solar power, can have a large physical footprint relative to energy produced; they constitute such a small part of the current energy mix in North America that their aggregate energy sprawl impact at present is modest but growing. Because wind turbines require minimum spacing distances to maximize wind energy capture, the physical footprint of wind power is extensive but can be mitigated, whereas decapitated mountains in Appalachia sacrificed for surface coal mining will never grow back. Siting wind turbines in existing agricultural landscapes need not fragment any additional wildlife habitat. Putting solar arrays on rooftops, parking lots, and urban brownfields need not contribute to energy sprawl at all while generating significant energy close to where it is needed, eliminating the sprawl precipitated by new transmission lines.

Devoting land to growing feedstock for liquid biofuels, or growing biomass for generating electricity, augurs the greatest potential energy sprawl of the major energy alternatives under discussion. The energy density of these fuels is low and the amount of land that must be effectively industrialized, even for relatively small quantities of biofuels or biomass-derived electricity, is massive. In the end, the most effective strategy for fighting energy sprawl is to reduce energy consumption.

VISUAL BLIGHT

The rampant air and water pollution resulting from fossil fuel use has garnered considerable attention in recent years, with landmark studies on the human health effects and other costs of coal burning, and alarming accounts of declining air quality in gas-and-oil-drilling boomtowns. Toxic accidents such as the 2010 Deepwater Horizon oil rig blowout in the Gulf of Mexico, and the massive 2008 spill of coal-combustion waste into the Clinch and Emory rivers in Tennessee, temporarily galvanize public attention before fading from the news. But the everyday nicks and cuts caused by energy development that mar the beauty and health of the Earth's ecosystems do not make headlines.

The creeping cancer of visual pollution is difficult to quantify but everywhere apparent. From strip mines and power lines to oil spills and sprawling wind power developments, energy-related visual blight is rampant. It is no accident that the great conservationist Aldo Leopold included aesthetics in his oft-quoted summation of the land ethic: "A thing is right when it tends to support the integrity, stability, and beauty of the biotic community. It is wrong when it tends otherwise." Ecological integrity and beauty are connected. Wild, intact landscapes are intrinsically beautiful. Degraded, despoiled landscapes are ugly.

Beauty matters greatly to human health and happiness but is frequently discounted. Conservationists fighting industrial-scale wind projects because of their visual impacts are mocked by wind power boosters as being overly concerned about "scenery." This is an old refrain; legendary congressman Joe Cannon, the Speaker of the U.S. House of Representatives who frequently opposed President Teddy Roosevelt's progressive conservation policies, once quipped, "Not one cent for scenery." Boosters of economic growth have been echoing that ignorant sentiment for the past century.

An energy economy that truly supported the health of the land community would not degrade beauty but foster it. It would promote ecological integrity on land and sea. A central criterion by which to measure any proposed energy development would be its contribution to visual pollution.

CLIMATE CHANGE

The mechanism by which certain gases, especially carbon dioxide and methane, hold additional heat in Earth's atmosphere that would otherwise be radiated off the planet's surface has been known since the nineteenth century. Global "greenhouse gas" pollution has been increasing significantly since the dawn of the Industrial Revolution, and the current level of carbon dioxide in the atmosphere is higher than at any time in at least eight hundred thousand years.

Unfortunately for the community of life that has evolved within the climatic conditions of the past few hundred thousand years—a community that includes humans—the consequences of these human-caused disruptions to the planet's climate are likely to be grim. Record droughts and floods are already being felt around the globe and are projected to worsen in coming decades, even if a miraculous cessation of greenhouse gas emissions were to occur. The authoritative (and conservative) Intergovernmental Panel on Climate Change, which reflects the view of the world's leading climate scientists, has documented various impacts of anthropogenic climate change including rising sea level, diminishing sea ice and melting glaciers, and increased ocean acidification. Unprecedented heat waves from Pakistan to Moscow and unusually severe storms in North America—the kinds of events expected as a consequence of global warming—have affected many millions of people. And these dramatic indicators of climate chaos are linked to an increased global temperature of a mere 1.4°F over the past century. Many experts believe that these changes are merely a hint of what is to come as the global climate reaches a tipping point, where feedback loops such as methane release from melting permafrost and reduced reflectivity from disappearing Arctic ice cause runaway overheating. Climate chaos threatens to amplify the existing species extinction crisis, and to overwhelm the world's economies with increasing costs and declining productivity.

Although a global movement of citizen activists concerned about climate change has emerged, governmental action at all levels has been inadequate to confront the magnitude of the problem. The fastest-ever growth in carbon dioxide emissions was recorded in 2010. Year after year, global greenhouse gas pollution records are broken as the surging economies of India and China strive to catch up with the old-guard polluters. The challenge is formidable because the interconnected factors of overpopulation, economic growth, development philosophy, and the current energy economy must all be addressed if there is any hope to forestall the worst effects of climate change.

ENERGY CONSERVATION

Americans comprise a mere 4.5 percent of the global population but consume about 20 percent of energy output annually. The silver lining in that dark cloud of profligacy is that the present energy economy is so wasteful and inefficient that it offers extraordinary potential for energy savings through conservation and efficiency efforts.

In common parlance *conservation* means using less energy, which can be accomplished through *curtailment* (turning off the lights when you leave the room) and *efficiency* (leaving the lights on but replacing old bulbs with high-efficiency LED bulbs).

In the 1970s, following the OPEC oil embargo when fuel prices spiked, America briefly got interested in energy conservation. Jimmy Carter donned a sweater on national TV and put solar panels on the While House roof. But Ronald Reagan took the solar panels down, and two decades of low fossil fuel prices spawned living patterns that used vast amounts of energy, including suburban and exurban housing developments serviced by fleets of SUVs. In hindsight, these choices look foolish—although the present era of high oil prices hasn't yet prompted a major political party's national convention to erupt in cheers of "Save, Baby, Save."

Supply-side boosterism has long kept conservation the ugly stepchild of energy policy, but as energy becomes less affordable, Americans may again be ready to consider the compelling reasons why conservation should become the centerpiece of national energy policy. Those reasons include conservation's contribution to increasing national security, as well as to creating jobs, stimulating innovation, saving money, reducing carbon emissions, and lessening the energy economy's impacts on wild nature. Numerous studies have analyzed various sectors of the energy economy and outlined the huge potential for intelligent conservation programs to reduce energy consumption while maintaining, and often improving, quality of life.

EFFICIENCY

Global energy consumption has spiked along with population growth and increasing affluence, but there are vast disparities in energy use between people living in the leading industrialized countries and those in less-industrialized nations. And even in the overdeveloped nations, significant disparity exists; the average American, for instance, uses twice as much energy as a typical European to achieve an equivalent lifestyle. That ratio can be viewed as dangerously unsustainable, or as an opportunity to begin picking the low-hanging fruit offered by energy conservation programs.

With so much waste and inefficiency built into its current energy system, the United States has tremendous potential to harvest "negawatts" through efficiency and curtailment. A 2009 report by consulting giant McKinsey and Company outlined readily achievable efficiency measures for the non-transportation part of U.S. energy economy that, if aggressively implemented over the next decade, could reduce projected 2020 energy demand by roughly 23 percent—saving more than $1.2 trillion in energy costs. Pioneering efficiency advocates at the Rocky Mountain Institute have developed an even more ambitious agenda for massive potential savings, in energy and dollars, of efficiency. RMI's Reinventing Fire blueprint outlines a path to a U.S. economy larger by 158 percent in the year 2050 that would require no oil, coal, or nuclear energy, and would save trillions of dollars, thus enhancing business profitability.

California, which has the world's eighth largest economy, is an often-cited example of real-world success in conservation programs that have reduced load growth over time. Following the energy crisis in the 1970s, the state instituted a suite of policies to promote efficiency, including letting electricity costs rise significantly. In the following decades, per capita electricity use in the state remained essentially flat, while the national average increased some 50 percent. These measures helped decouple growth in electricity consumption from population and economic growth, although aggregate electricity use did increase by an average of 1.6 percent annually in the twenty-five years following 1980. Growth at even that modest rate means a doubling of energy use in less than forty-four years, which is not sustainable in a world bumping up against resource constraints.

In essence, the California experience shows what can happen when people in a very wasteful system get serious about harvesting low-hanging fruit. The initial efficiency gains are the easiest, but then the law of diminishing returns kicks in. As the waste is worked out of the system and it nears maximum efficiency, there is less fruit available to pick and it becomes more difficult to reach. At its maximum efficiency, a system is right at the edge of its breaking point. The overarching reality is that continuous growth on a finite planet is physically impossible. Efficiency programs have tremendous potential and are extremely cost effective, but ultimately they cannot keep up if exponential population and economic growth continues.

CURTAILMENT WRIT LARGE

Ask a roomful of energy experts about the future and you'll get a wide range of opinions: Peaking oil production will necessitate a shift to less energy dense fuels and cause energy costs to spike. Resource scarcity will precipitate additional conflict between nations. Climate change will cause mass dislocations and civilizational collapse. Or, on the opposite side of the spectrum, human creativity and the power of the marketplace will unleash innovations that will provide plentiful, low-cost, low-carbon energy.

The precise future is unknown, but the present is certain: Our current energy economy is destructive to nature and dangerous to democratic institutions, community life, and human health. It is a toxic system that requires fundamental reform. The political, philosophical, economic, and physical barriers to rebuilding the energy economy make the task difficult. But it is achievable.

Barring some breakthrough in energy technology or a global pandemic that decimates human numbers (two scenarios that are possible but unlikely), over the next century there will be less energy per capita available than during the last. Thus, a policy of curtailment—ending frivolous and wasteful uses of energy, deploying existing renewable power technologies in the most effective manner, and ratcheting down fossil fuel use quickly and dramatically so that the planetary fever goes no higher—is the overarching task for our times.

Americans, who have become accustomed to the idea that anyone should be able to use as much energy as they want, whenever they want, for whatever purpose (and it should be cheap!), will face a different reality in an energy-constrained future. In a sane world, we would not blow the tops off mountains in Appalachia to keep coal-burning power plants belching pollution so that office towers can leave the lights on all night. From motorized paper-towel dispensers and illuminated, empty parking lots to the worst inefficiencies of suburban sprawl, there are worlds of energy-wasting products, activities, and living arrangements that can and should simply be abandoned. Curtailment achieved through outright abolition of energy-wasting machines or activities would be controversial. Nevertheless, in an energy-constrained world with a bad case of human-induced hemorrhagic fever, the sooner citizens voluntarily begin curtailment efforts, the more options remain open to transition toward a more durable, ecologically sustainable energy system.

Part Two

THE PREDICAMENT

INTRODUCTION

Energy, Nature, and the Eco-Social Crisis

A grounding in energy fundamentals—energy literacy—is necessary for any citizen wanting to participate in the crucial conversation about energy policy and the future of nature and people. But it is not sufficient. To answer such questions as: Where did the present energy economy come from? Where is it going? What are the unstated assumptions about "progress" that inform energy policy?—it is also necessary to consider the history and ideas that undergird the entire globalized, industrial growth culture.

Without considering the philosophical foundations of that culture, there is no way to articulate a deep systemic critique that may orient society toward fundamental change which addresses root causes, rather than superficial reforms that merely ameliorate symptoms. Developing a systemic critique naturally leads one to name the various drivers of the global eco-social crisis, the great ecological and social unraveling that defines our times. Growth-based economics, mass consumerism, globalization, rapacious corporate profit, unintended negative effects of technology, and the human population explosion will be among the identified agents of destruction. Rising social inequity, climate change, loss of beauty, and, most fundamentally, the pervasive destruction of natural habitats across the globe, are among the consequences.

While problems affecting the health and welfare of human beings capture most attention and funding, it is the health and welfare of nature that supports the entire community of life—and also determines whether any particular human culture is sustainable over the long term. The present mass extinction event, a contraction in life's diversity unprecedented in the past 65 million years, should be the clearest marker of all that humanity is on the wrong trajectory. Human-caused extinction of our fellow species is not merely a moral blight on *Homo sapiens* but also a dire concern: All human economies are a subset of nature's economy, and thus the fate of people is inextricably tied to the fate of nature.

The eco-social crisis may have many drivers, but it is the *idea* that humans are separate from, and sovereign over, the rest of life that is at the heart of the problem. It may well be that the capacity and constant desire to innovate is the most distinctive, and ultimately problematic, facet of human nature. It remains to be seen whether that capacity can be reined in, harnessed to a grounding philosophy of *ecocentrism* that stresses the intrinsic value of all life. But if so, that return to ways of thinking that shaped human cultural development for the vast majority of our history as a species would do far more to refashion the energy economy than any new policy or technology.

Because the spectacular windfall of fossil energy that allowed the Industrial Revolution to spread the trappings of modernity around the globe has seemed so limitless, it is bound to be a difficult transition to return our thinking to one bounded by natural limits, humility, and neighborliness. The first step in that challenging but possible transition requires looking forthrightly at humanity's current predicament, and the worldview behind it.

FIVE CARBON POOLS

WES JACKSON

THE INVENTION OF *agriculture ten millennia ago was the first step toward the current problem of climate change. Humans then began a way of life that would exploit the first of five relatively nonrenewable pools of energy-rich carbon—soil. Trees, coal, oil, and natural gas would follow as additional pools to rob from. We are the first species in this multibillion-year journey of life on Earth that will have to practice restraint after years of reckless use of the five carbon pools. We have to stop deficit spending of the ecosphere and reduce our numbers if we hope to prevent widespread sociopolitical upheaval.*

❖

THE FIRST POOL

Our very being, the physical and cognitive attributes of *Homo sapiens*, was shaped by a seamless series of changing ecosystems embedded within an ever-changing ecosphere over hundreds of millions of years. The planet's ability to support humans into a distant future was not on the line. The context of our livelihood kept our numbers more or less in check. Diseases killed us. Predators ate us. Sometimes we starved. The context that had shaped us was the context within which we lived. Apparently we had been eating grains but not improving them for centuries. But something happened some ten millennia ago called the Agricultural Revolution. It also became a treadmill. It happened first in one of these ecosystems, most likely in the land to the east of the Mediterranean, but soon spread. Hunter–gatherers initiated what would be recognized later as a break with nature, a split. This new way of being began our escape from gathering and hunting as a way of life. To set the record straight, Eden was no garden and our escape only partial. Where we planted our crops, we reduced the diversity of the biota. The landscape simplified by agriculture locked our ancestors into a life of "thistles, thorns, and sweat of brow." We became a species out of context. It has been said that if we were meant to be agriculturists, we would have had longer arms.

No matter how unpleasant this agricultural work may have been, the food calories increased. Our numbers rose; more mouths needed to be fed. No matter that they disliked thistles and thorns and sweat of brow, our ancestors loved their children and their own lives, and so they kept doing it. They had to eat. Some gave up agriculture when they had the chance. The introduction of the horse by the Spanish allowed some of the Native Americans to return to hunting and gathering, for a short while. Eventually the draft animals, especially the ox and the horse, were domesticated. These creatures used the stored sunlight of a grass, shrub, or tree leaf and transferred it to the muscle used to pull a plow or bear a load. They became "beasts of burden."

This step onto the agriculture treadmill was the first toward the current and looming problem of climate change. It was in that time that humans began a way of life that would exploit the first of five relatively nonrenewable pools of energy-rich carbon—soil. Trees, coal, oil, and natural gas would follow as additional pools to rob from. Our crops and we—both of us—were beneficiaries of the energy released as nutrients stored in the carbon compounds in the soil now became available. It was agriculture that featured annuals in monoculture instead of perennials in mixtures where the split with nature began. And so it was at this moment that the carbon compounds of the soil were exposed to more rapid oxidation. Carbon dioxide headed for the atmosphere, and the nutrients formerly bound up in those organic compounds—nutrients such as phosphorus and potassium—were now available for uptake by our annual crop plants. So, this wasn't really a use of the energy-rich carbon in the sense that we were after the energy stored in the carbon molecule. Rather, the breaking of the carbon compound at work in the soil was a consequence of agriculture. With agriculture, the soils that had once safely absorbed the footsteps of the Paleolithic gatherers and hunters and their food supply lay vulnerable. The hoe, along with the power to domesticate plants into crops and wild animals into livestock, turned these people into the most important revolutionaries our species has ever known. They plunged ahead in this new way of life, repeatedly modifying their agrarian technique as they went.

How many were aware they were at the forefront of a way of life dependent on deficit spending of the Earth's capital? Certainly long before the advent of writing, humans must have understood that till agriculture not only simplifies the landscape but also compromises soil quality and plant fertility. Even so, the reality informed by the immediate reigned. People needed food. Energy-rich carbon molecules were the workhorses in the soil accommodating a diversity of species. The seeds from annual crop monocultures would feed the tribe. The energy-rich carbon in the grains provided these tribes with a more reliable and abundant food supply and, therefore, made possible the beginning of civilization. Eventually the descendants of these farmers had the tools necessary to expand the scale of shrub and tree harvest. Now the agriculturists could more aggressively exploit the second nonrenewable pool—forests.

THE SECOND POOL

Five thousand or so years passed. It is easy to imagine that as the agriculturists wandered through the forests, their curious minds saw that they could cut down the forests to purify ores. This led to the creation some five thousand years ago of first the Bronze Age and then the Iron Age, and led to a further distancing of nature. But soon this second pool of energy-rich carbon was on its way to being used up beyond local replacement levels. This second use of carbon—deforestation—became, unambiguously, a mining operation. And it came on fast. And so the forests went down as the soils were eroding, first in the Middle East and later in Europe and Asia. And so it went for millennia, relentlessly, until recently.

THE THIRD POOL

Only one-quarter of one millennium ago, the third pool—coal—was opened on a large scale with the launching of the Industrial Revolution in 1750. But already by 1700, England's forests were mostly gone to heat the pig iron. The Brits then took their ore to Ireland, where forests were still abundant, to purify the metal. The stock of the second pool of energy-rich carbon, the forests, had been so depleted that this third pool must have gladdened the hearts of those who would exploit it. Coal reduced the pressure on the forests only slightly, for after the defeat of the Spanish Armada, it cost England its forests to rule the waves for the next three hundred years.

The availability of coal, this third pool, provided a quantum leap in our ability to accomplish more work in a shorter period of time. The density of energy stored in a pound of coal is far greater than the density in a pound of wood. The accessibility and breakability of coal sponsored countless hopes, dreams, and aspirations of the British Empire. However, the colonialism those carbon pools made possible also destroyed local cultural and ecological arrangements that will be, at best, slow to replace in a Sun-powered world.

It seems inevitable now that Neolithic farmers would move from a Stone Age and on to a Bronze Age, and later, an Iron Age. Similarly, given the energy density of coal, it also seems inevitable now that a steam engine would be built to accelerate the Industrial Revolution. Without soil carbon, forests, and coal, it seems doubtful that the British Empire would have had the slack in 1831 to send a young Charles Darwin on his famous voyage around the world. And once home, he was given the leisure to investigate his collections, pore over his journals, exchange letters with contemporaries, converse with his scientific peers, and finally, in 1859, have *On the Origin of Species* appear in London bookstores.

THE FOURTH POOL

The year 1859 was an auspicious one, beyond Darwin's publication. It was also the year of the first oil well—Colonel Edwin Drake's oil well in western Pennsylvania—and the opening of the fourth pool of energy-rich carbon, oil. Cut a tree and you have to either chop or saw it into usable chunks. Coal you have to break up. But oil is a portable liquid fuel transferable in a pipe, a perfect product of the Iron Age.

The year 1859 was also when the ardent abolitionist John Brown was hanged at Harpers Ferry, a reality more than coincidental. In some respects, John Brown, beyond believing in the absolute equality of blacks and whites, stands alone in his time. His fervor would have received little traction had not the numbers of abolitionists been growing in the industrial North. The South had coal, of course, but not as much. It was a more agrarian society. Northern supporters, who were more profligate carbon-pool users, could afford to be more self-righteous than the more agrarian, less coal-using, slaveholding South. Leisure often makes virtue easier.

THE FIFTH POOL

Natural gas has been available in some form of use back to the times of the ancient Greeks. But it did not become a manageable pool as a major power source until after coal began to be used. We count it as the fifth pool and likely the last major pool. Other minor pools may follow, such as the lower-quality tar sands and shale oil—both energy- and water-intensive for their extraction—which are in the early stages of being exploited. Over the last half century, we have used natural gas as a feedstock to make nitrogen fertilizer, which we apply to our fields to provide us a bountiful food supply while creating dead zones in our oceans. This technology, called the Haber-Bosch process, was developed in the first decade of the twentieth century by two Germans, Fritz Haber and Carl Bosch. Vaclav Smil, a resource scholar at the University of Manitoba, has called it "the most important invention of the twentieth century." Without it, Smil says, 40 percent of humanity would not be here. This is certainly a true enough statement given the reality of our cattle, pig, and chicken welfare programs.

When we were gatherers and hunters, the ecosystem kept us in check. But since the advent of agriculture, we have forced the landscape to meet our expectations, and we have been centered on this way of life. We plow. We cut forests. We mine coal. We drill for oil and natural gas. We want the stored sunlight the oxygen helps release. The oxygen that enters our lungs to oxidize energy-rich carbon molecules in our cells is internal combustion—not too dissimilar to the oxygen that enters the air intake of an automobile and, with the aid of a spark, releases the energy to power a bulldozer or to run a car idling in a traffic jam.

We relentlessly rearrange the five carbon pools to get *more* energy or more useful materials. Internal combustion is the name of the game. We reorder our landscapes and industrial machinery to keep our economic enterprises (and ourselves) going, all the while depleting the stocks of nonrenewable energy-rich carbon. We are like bacteria on a Petri dish with sugar.

So here we are, the first species in this multibillion-year journey of life on Earth that will have to practice restraint after years of reckless use of the five carbon pools. None of our ancestors had to face this reality. We are living in the most important and challenging moment in the history of *Homo sapiens*, more important than any of our wars, more important than our walk out of Africa. More important than any of our conceptual revolutions. We have to consciously practice restraint to end our "use it till it's gone" way of life. We have to stop deficit spending of the ecosphere and reduce our numbers if we hope to prevent widespread sociopolitical upheaval.

FAUSTIAN ECONOMICS

Hell Hath No Limits

WENDELL BERRY

WE HAVE FOUNDED *our present society upon delusional assumptions of limitlessness. Our national faith is a sort of autistic industrialism. This necessarily leads to limitless violence, waste, war, and destruction. Our great need now is for sciences and technologies of limits. The limits would be the accepted contexts of places, communities, and neighborhoods, both natural and human. To make our economic landscapes sustainably and abundantly productive, we must maintain in them a living formal complexity something like that of natural ecosystems. We can do this only by raising to the highest level our mastery of the arts of agriculture, animal husbandry, forestry, and, ultimately, the art of living.*

❖

The general reaction to the apparent end of the era of cheap fossil fuel, as to other readily foreseeable curtailments, has been to delay any sort of reckoning. The strategies of delay, so far, have been a sort of willed oblivion, or visions of large profits to the manufacturers of such "biofuels" as ethanol from corn or switchgrass, or the familiar unscientific faith that "science will find an answer." The dominant response, in short, is a dogged belief that what we call the American Way of Life will prove somehow indestructible. We will keep on consuming, spending, wasting, and driving, as before, at any cost to anything and everybody but ourselves.

This belief was always indefensible—the real names of global warming are Waste and Greed—and by now it is manifestly foolish. But foolishness on this scale looks disturbingly like a sort of national insanity. We seem to have come to a collective delusion of grandeur, insisting that all of us are "free" to be as conspicuously greedy and wasteful as the most corrupt of kings and queens. (Perhaps by devoting more and more of our already abused cropland to fuel production we will at last cure ourselves of obesity and become fashionably skeletal, hungry but—thank God!—still driving.)

The problem with us is not only prodigal extravagance but also an assumed limitlessness. We have obscured the issue by refusing to see that limitlessness is a godly trait. We have insistently, and with relief, defined ourselves as animals or as "higher animals." But to define ourselves as animals, given our specifically human powers and desires, is to define ourselves as *limitless* animals—which of course is a contradiction in terms. Any definition is a limit, which is why the God of Exodus refuses to define Himself: "I am that I am."

Even so, that we have founded our present society upon delusional assumptions of limitlessness is easy enough to demonstrate. A recent "summit" in Louisville, Kentucky, was entitled "Unbridled Energy: The Industrialization of Kentucky's Energy Resources." Its subjects were "clean-coal generation, biofuels, and other cutting-edge applications," the conversion of coal to "liquid fuels," and the likelihood that all this will be "environmentally friendly." These hopes, which "can create jobs and boost the nation's security," are to be supported by government "loan guarantees...investment tax credits and other tax breaks." Such talk we recognize as completely conventional. It is, in fact, a tissue of clichés that is now the common tongue of promoters, politicians, and journalists. This language does not allow for any computation or speculation as to the *net* good of anything proposed. The entire contraption of "Unbridled Energy" is supported only by a rote optimism: "The United States has 250 billion tons of recoverable coal reserves—enough to last one hundred years even at double the current rate of consumption." We humans have inhabited the Earth for many thousands of years, and now we can look forward to surviving for another hundred by doubling our consumption of coal? *This* is national security? The world-ending fire of industrial fundamentalism may already be burning in our furnaces and engines, but if it will burn for a hundred more years, that will be fine. Surely it would be better to intend straightforwardly to contain the fire and eventually put it out! But once greed has been made an honorable motive, then you have an economy without limits. It has no place for temperance or thrift or the ecological law of return. It will do anything. It is monstrous by definition.

In keeping with our unrestrained consumptiveness, the commonly accepted basis of our economy is the supposed possibility of limitless growth, limitless wants, limitless wealth, limitless natural resources, limitless energy, and limitless debt. The idea of a limitless economy implies and requires a doctrine of general human limitlessness: *All* are entitled to pursue without limit whatever they conceive as desirable—a license that classifies the most exalted Christian capitalist with the lowliest pornographer.

This fantasy of limitlessness perhaps arose from the coincidence of the Industrial Revolution with the suddenly exploitable resources of the New World—though how the supposed limitlessness of resources can be reconciled with their exhaustion is not clear. Or perhaps it comes from the contrary apprehension of the world's "smallness," made possible by modern astronomy and high-speed transportation. Fear of the smallness of our world and its life may lead to a kind of claustrophobia and thence, with apparent reasonableness, to a desire for the "freedom" of limitlessness. But this desire, paradoxically, reduces everything. The life of this world *is* small to those who think it is, and the desire to enlarge it makes it smaller, and can reduce it finally to nothing.

However it came about, this credo of limitlessness clearly implies a principled wish not only for limitless possessions but also for limitless knowledge, limitless science, limitless technology, and limitless progress. And, necessarily, it must lead to limitless violence, waste, war, and destruction. That it should finally produce a crowning cult of political limitlessness is only a matter of mad logic.

The normalization of the doctrine of limitlessness has produced a sort of moral minimalism: the desire to be efficient at any cost, to be unencumbered by complexity. The minimization of neighborliness, respect, reverence, responsibility, accountability, and self-subordination —this is the culture of which our present leaders and heroes are the spoiled children.

Our national faith so far has been: "There's always more." Our true religion is a sort of autistic industrialism. People of intelligence and ability seem now to be genuinely embarrassed by any solution to any problem that does not involve high technology, a great expenditure of energy, or a big machine. Thus an X marked on a paper ballot no longer fulfills our idea of voting. One problem with this state of affairs is that the work now most needing to be done—that of neighborliness and caretaking—cannot be done by remote control with the greatest power on the largest scale.

A second problem is that the economic fantasy of limitlessness in a limited world calls fearfully into question the value of our monetary wealth, which does not reliably stand for the real wealth of land, resources, and workmanship but instead wastes and depletes it.

That human limitlessness is a fantasy means, obviously, that its life expectancy is limited. There is now a growing perception, and not just among a few experts, that we are entering a time of inescapable limits. We are not likely to be granted another world to plunder in compensation for our pillage of this one. Nor are we likely to believe much longer in our ability to outsmart, by means of science and technology, our economic stupidity. The hope that we can cure the ills of industrialism by the homeopathy of more technology seems at last to be losing status. We are, in short, coming under pressure to understand ourselves as limited creatures in a limited world.

This constraint, however, is not the condemnation it may seem. On the contrary, it returns us to our real condition and to our human heritage, from which our self-definition as limitless animals has for too long cut us off. Every cultural and religious tradition that I know about, while fully acknowledging our animal nature, defines us specifically as *humans*—that is, as animals (if the word still applies) capable of living not only within natural limits but also within cultural limits, self-imposed. As earthly creatures, we live, because we must, within natural limits, which we may describe by such names as "Earth" or "ecosystem" or "watershed" or "place." But as humans, we may elect to respond to this necessary placement by the self-restraints implied in neighborliness, stewardship, thrift, temperance, generosity, care, kindness, friendship, loyalty, and love.

In our limitless selfishness, we have tried to define "freedom," for example, as an escape from all restraint. But, as my friend Bert Hornback has explained in his book *The Wisdom in Words*, "free" is etymologically related to "friend." These words come from the same Indo-European root, which carries the sense of "dear" or "beloved." We set our friends free by our love for them, with the implied restraints of faithfulness or loyalty. And this suggests that our "identity" is located not in the impulse of selfhood but in deliberately maintained connections.

Thinking of our predicament has sent me back again to Christopher Marlowe's *Tragical History of Doctor Faustus*. This is a play of the Renaissance; Faustus, a man of learning, longs to possess "all Nature's treasury," to

"Ransack the ocean ... / And search all corners of the new-found world...."
To assuage his thirst for knowledge and power, he deeds his soul to Lucifer,
receiving in compensation for twenty-four years the services of the sub-devil
Mephistophilis, nominally Faustus's slave but in fact his master. Having the
subject of limitlessness in mind, I was astonished on this reading to come
upon Mephistophilis's description of hell. When Faustus asks, "How comes
it then that thou art out of hell?" Mephistophilis replies, "Why, this is hell,
nor am I out of it." And a few pages later he explains:

> Hell hath no limits, nor is circumscribed.
> In one self place, but where we [the damned] are is hell,
> And where hell is must we ever be.

For those who reject heaven, hell is everywhere, and thus is limitless.
For them, even the thought of heaven is hell.

It is only appropriate, then, that Mephistophilis rejects any conventional limit: "Tut, Faustus, marriage is but a ceremonial toy. If thou lovest
me, think no more of it." Continuing this theme, for Faustus's pleasure the
devils present a sort of pageant of the seven deadly sins, three of which—
Pride, Wrath, and Gluttony—describe themselves as orphans, disdaining
the restraints of parental or filial love.

Seventy or so years later, and with the issue of the human definition more than ever in doubt, John Milton in Book VII of *Paradise Lost*
returns again to a consideration of our urge to know. To Adam's request
to be told the story of creation, the "affable Archangel" Raphael agrees
"to answer thy desire / Of knowledge *within bounds* [my emphasis]...."
explaining that

> Knowledge is as food, and needs no less
> Her temperance over appetite, to know
> In measure what the mind may well contain;
> Oppresses else with surfeit, and soon turns
> Wisdom to folly, as nourishment to wind.

Raphael is saying, with angelic circumlocution, that knowledge
without wisdom, limitless knowledge, is not worth a fart; he is not a
humorless archangel. But he also is saying that knowledge without measure, knowledge that the human mind cannot appropriately use, is mor-
tally dangerous.

I am well aware of what I risk in bringing this language of religion into what is normally a scientific discussion. I do so because I doubt that we can define our present problems adequately, let alone solve them, without some recourse to our cultural heritage. We are, after all, trying now to deal with the failure of scientists, technicians, and politicians to "think up" a version of human continuance that is economically probable and ecologically responsible, or perhaps even imaginable. If we go back into our tradition, we are going to find a concern with religion, which at a minimum shatters the selfish context of the individual life, and thus forces a consideration of what human beings are and ought to be.

This concern persists at least as late as our Declaration of Independence, which holds as "self-evident, that all men are created equal, that they are endowed by their Creator with certain unalienable Rights." Thus among our political roots we have still our old preoccupation with our definition as humans, which in the Declaration is wisely assigned to our Creator; our rights and the rights of all humans are not granted by any human government but are innate, belonging to us by birth. This insistence comes not from the fear of death or even extinction but from the ancient fear that in order to survive we might become inhuman or monstrous.

And so our cultural tradition is in large part the record of our continuing effort to understand ourselves as beings specifically human: to say that, as humans, we must do certain things and we must not do certain things. We must have limits or we will cease to exist as humans; perhaps we will cease to exist, period. At times, for example, some of us humans have thought that human beings, properly so called, did not make war against civilian populations, or hold prisoners without a fair trial, or use torture for any reason.

Some of us would-be humans have thought too that we should not be free at anybody else's expense. And yet in the phrase "free market," the word "free" has come to mean unlimited economic power for some, with the necessary consequence of economic powerlessness for others. Several years ago, after I had spoken at a meeting, two earnest and obviously troubled young veterinarians approached me with a question: How could they practice veterinary medicine without serious economic damage to the farmers who were their clients? Underlying their question was the fact that for a long time veterinary help for a sheep or a pig has been likely to cost more than the animal is worth. I had to answer that, in my opinion, so long as their practice relied heavily on selling patented drugs, they had no choice, since the market for medicinal drugs was entirely controlled by the drug companies, whereas most farmers had no control at all over the market for agricultural products. My questioners were asking in effect if a predatory economy can have a beneficent result. The answer too often is no. And that is because there is an absolute discontinuity between the economy of the seller of medicines and the economy of the buyer, as there is in the health industry as a whole. The drug industry is interested in the survival of patients, we have to suppose, because surviving patients will continue to consume drugs.

Now let us consider a contrary example. Recently, at another meeting, I talked for some time with an elderly, and some would say an old-fashioned, farmer from Nebraska. Unable to farm any longer himself, he had rented his land to a younger farmer on the basis of what he called "crop share" instead of a price paid or owed in advance. Thus, as the old farmer said of his renter, "If he has a good year, I have a good year. If he has a bad year, I have a bad one." This is what I would call community economics. It is a sharing of fate. It assures an economic continuity and a common interest between the two partners to the trade. This is as far as possible from the economy in which the young veterinarians were caught, in which the powerful are limitlessly "free" to trade, to the disadvantage, and ultimately the ruin, of the powerless.

It is this economy of community destruction that, wittingly or unwittingly, most scientists and technicians have served for the past two hundred years. These scientists and technicians have justified themselves by the proposition that they are the vanguard of progress, enlarging human knowledge and power, and thus they have romanticized both themselves and the predatory enterprises that they have served.

As a consequence, our great need now is for sciences and technologies of limits, of domesticity, of what Wes Jackson of The Land Institute in Salina, Kansas, has called "homecoming." These would be specifically human sciences and technologies, working, as the best humans always have worked, within self-imposed limits. The limits would be the accepted contexts of places, communities, and neighborhoods, both natural and human.

I know that the idea of such limitations will horrify some people, maybe most people, for we have long encouraged ourselves to feel at home on "the cutting edges" of knowledge and power or on some "frontier" of human experience. But I know too that we are talking now in the presence of much evidence that improvement by outward expansion may no longer be a good idea, if it ever was. It was not a good idea for the farmers who "leveraged" secure acreage to buy more during the 1970s. It has proved tragically to be a bad idea in a number of recent wars. If it is a good idea in the form of corporate gigantism, then we must ask, For whom? Faustus, who wants all knowledge and all the world for himself, is a man supremely lonely and finally doomed. I don't think Marlowe was kidding. I don't think Satan is kidding when he says in *Paradise Lost*, "Myself am Hell."

If the idea of appropriate limitation seems unacceptable to us, that may be because, like Marlowe's Faustus and Milton's Satan, we confuse limits with confinement. But that, as I think Marlowe and Milton and others were trying to tell us, is a great and potentially a fatal mistake. Satan's fault, as Milton understood it and perhaps with some sympathy, was precisely that he could not tolerate his proper limitation; he could not subordinate himself to anything whatever. Faustus's error was his unwillingness to remain "Faustus, and a man." In our age of the world it is not rare to find writers, critics, and teachers of literature, as well as scientists and technicians, who regard Satan's and Faustus's defiance as salutary and heroic.

On the contrary, our human and earthly limits, properly understood, are not confinements but rather inducements to formal elaboration and elegance, to *fullness* of relationship and meaning. Perhaps our most serious cultural loss in recent centuries is the knowledge that some things, though limited, are inexhaustible. For example, an ecosystem, even that of a working forest or farm, so long as it remains ecologically intact, is inexhaustible. A small place, as I know from my own experience, can provide opportunities of work and learning, and a fund of beauty, solace, and pleasure—in addition to its difficulties—that cannot be exhausted in a lifetime or in generations.

To recover from our disease of limitlessness, we will have to give up the idea that we have a right to be godlike animals, that we are potentially omniscient and omnipotent, ready to discover "the secret of the universe." We will have to start over, with a different and much older premise: the naturalness and, for creatures of limited intelligence, the necessity, of limits. We must learn again to ask how we can make the most of what we are, what we have, what we have been given. If we always have a theoretically better substitute available from somebody or someplace else, we will never make the most of anything. It is hard to make the most of one life. If we each had two lives, we would not make much of either. Or as one of my best teachers said of people in general: "They'll never be worth a damn as long as they've got two choices."

To deal with the problems, which after all are inescapable, of living with limited intelligence in a limited world, I suggest that we may have to remove some of the emphasis we have lately placed on science and technology and have a new look at the arts. For an art does not propose to enlarge itself by limitless extension but rather to enrich itself within bounds that are accepted prior to the work.

It is the artists, not the scientists, who have dealt unremittingly with the problem of limits. A painting, however large, must finally be bounded by a frame or a wall. A composer or playwright must reckon, at a minimum, with the capacity of an audience to sit still and pay attention. A story, once begun, must end somewhere within the limits of the writer's and the reader's memory. And of course the arts characteristically impose limits that are artificial: the five acts of a play, or the fourteen lines of a sonnet. Within these limits artists achieve elaborations of pattern, of sustaining relationships of parts with one another and with the whole, that may be astonishingly complex. And probably most of us can name a painting, a piece of music, a poem or play or story that still grows in meaning and remains fresh after many years of familiarity.

We know by now that a natural ecosystem survives by the same sort of formal intricacy, ever-changing, inexhaustible, and no doubt finally unknowable. We know further that if we want to make our economic landscapes sustainably and abundantly productive, we must do so by maintaining in them a living formal complexity something like that of natural ecosystems. We can do this only by raising to the highest level our mastery of the arts of agriculture, animal husbandry, forestry, and, ultimately, the art of living.

It is true that insofar as scientific experiments must be conducted within carefully observed limits, scientists also are artists. But in science one experiment, whether it succeeds or fails, is logically followed by another in a theoretically infinite progression. According to the underlying myth of modern science, this progression is always replacing the smaller knowledge of the past with the larger knowledge of the present, which will be replaced by the yet larger knowledge of the future.

In the arts, by contrast, no limitless sequence of works is ever implied or looked for. No work of art is necessarily followed by a second work that is necessarily better. Given the methodologies of science, the law of gravity and the genome were bound to be discovered by somebody; the identity of the discoverer is incidental to the fact. But it appears that in the arts there are no second chances. We must assume that we had one chance each for *The Divine Comedy* and *King Lear*. If Dante and Shakespeare had died before they wrote those poems, nobody ever would have written them.

The same is true of our arts of land use, our economic arts, which are our arts of living. With these it is once-for-all. We will have no chance to redo our experiments with bad agriculture leading to soil loss. The Appalachian mountains and forests we have destroyed for coal are gone forever. It is now and forevermore too late to use thriftily the first half of the world's supply of petroleum. In the art of living we can only start again with what remains. And so, in confronting the phenomenon of "peak oil," we are really confronting the end of our customary delusion of "more." Whichever way we turn, from now on, we are going to find a limit beyond which there will be no more. To hit these limits at top speed is not a rational choice. To start slowing down, with the idea of avoiding catastrophe, *is* a rational choice, and a viable one if we can recover the necessary political sanity. Of course it makes sense to consider alternative energy sources, provided *they* make sense. But also we will have to reexamine the economic structures of our lives, and conform them to the tolerances and limits of our earthly places. Where there is no more, our one choice is to make the most and the best of what we have.

LIFE-AFFIRMING BEAUTY

SANDRA B. LUBARSKY

THERE IS A relationship between sustainability and beauty. Beauty is our way of describing encounters with life-affirmative patterns of relationships. The current energy economy produces ugliness, a form of violence against the world and its creatures, which have aesthetic worth and intrinsic value. Does our inability to speak forthrightly about ugliness make us unintentionally complicit with this ongoing obliteration?

❖

Is it possible to look at a strip-mined slice of the Appalachians and demur from a judgment of ugliness, insisting that "beauty is in the eye of the beholder" and complaints against mountain ruin and valley spoil lack objective measure? Or do we deceive our senses when we refuse to enter into an aesthetic judgment on, say, the drill pads pockmarking Wyoming's upper Green River Valley, a landscape being desecrated in pursuit of boundless energy? Can we treat the obliteration of a mountainside or the disfigurement of land that has been road-carved and pipelined as coolly as a gallery artifact, amused that someone, somewhere must find these reshaped landscapes appealing? Or does our inability to speak publicly about ugliness make us unintentionally complicit with this ongoing obliteration?

The word "ugly" is derived from the Old Norse word *ugga*, to dread. It is to this original meaning that we must return for a word that can endure the weight of environmental desecration. "Dread" implies "fear," both words gaining their emotional strength from a threat of violence. The word "repugnant," a synonym of dread, wears its relation to violence in its root, *pugnare*, to fight, bringing to mind a foul, sickening fist. Etymologically, there is a direct line between ugliness and dread, dread and violence, and this line recovers the relation of ugliness to things that blunt or reduce life.

It is a mistake to think that the judgment, "ugly," is merely an expression of one person's opinion with little basis in shared, public experience. Though cultural habit plays a part in aesthetic judgment—we live our lives, after all, in a fine mesh of expectations and sensitivities—patterns that are brutally life-denying are repudiated cross-culturally. The calculated torture of a living being is one such pattern. The despoilment of nature is another. Their common measure is violence.

It is another mistake to think of ugliness as simply a synonym for "displeasure" or beauty as a synonym for "pleasure." When we do, we continue the modern conceit that the world's worth is a matter of human judgment. Such thinking turns nonhuman life into objects for our enjoyment—and for our use and abuse. "What good is a mountain just to have a mountain?" asked a representative of the West Virginia Coal Association in a public debate over whether Blair Mountain would remain a mountain or become a slagheap. Clearly, one of the best ways to objectify the world is to deny it a value of its own and to think of its worth only in terms of what we like or dislike.

All the many forms of life possess their own aesthetic worth; they do not depend on our pleasure or displeasure for confirmation of their value. When we think of value as intrinsic to all living beings, objectively present in the world, the worth of the world is secured in a way that exceeds human favor or disfavor. Whatever pleasure or disgust we register is in response to the objective factuality of beauty or its absence. A reintroduction of value into the world deposes materialism and its claim that the really real things that make up the world are senseless and valueless units of some quantifiable measure. We begin to see the world as filled with value and with feeling beings whose value contributes to the richness of experience. To know this is to realize that displeasure is too small a word to carry the freight of disorder and abuse barreling down on our natural systems.

Because the world is permeated with value, there is no neutral landscape; the world is never a blank slate, a neutral palette, a valueless fact. Between beauty and ugliness is a zero-sum relationship. When beauty is lost, it is replaced by ugliness. Just as evil is not simply the absence of good but a very potent force for attenuating life, so ugliness is not simply the absence of beauty but a devitalizing force, disruptive of the delicate adjustments of life. When particular forms of beauty are gone, they are as lost as the passenger pigeon or the great auk. To mourn the loss of life is also to mourn the loss of beauty.

There is a relationship between beauty and life, ugliness and the diminishment of life. Beauty is a marker of vitality; it is our way of describing encounters with life-affirmative patterns of relationships. To think of beauty in this way is to free it from the narrow constraints of either art or advertising. Beauty is one of the plain facts of experience, a part of our deep, evolutionary memory, kept alive in our daily consort with the world. We associate beauty with life-affirming conditions, arising as part of the miracle of relations that constitute the structure of life. A healthy ecosystem is predicated on countless fine-tuned adjustments. In biological terms, these adjustments are to a range of variables that includes such things as temperature, humidity, genetic variation, and species distribution. In aesthetic terms, millions of life-patterns are interwoven, each pattern enhanced by the others and contributing to the enrichment of all scales of life. What we have come to identify as beautiful are those patterns of adjustment that both enable and express life.

There is a relationship between sustainability and beauty. We need to name that relationship in order to give emotional honesty to our lives. A world in decline is a world that has become less beautiful. If we are to understand the widespread destructiveness of our current energy-hungry culture, we must attend to the loss of beauty and the increase of ugliness. Our inability to admit this and to address it will keep us imprisoned in the materialistic framework that bears much responsibility for the deterioration of the planet.

The poet Li Young Li once said, "When you become an artist, the material speaks back to you." As with poetry, so with beauty in general: Attend to beauty and the world speaks back, making its beauty evident or its lack palpable. In its presence and its absence, we are called to language and its power to make us responsible to the world. To admit neither to beauty nor ugliness is to live delinquently in the world. In speechlessness is acquiescence to the status quo.

It is all of a piece: the need to move to a worldview in which value is in the world and not only in the mind of the beholder, the need to reconnect beauty with life and our efforts to sustain the world, and the need to restore the language of beauty so that we can make judgments of value in the public

a team of scientists concluded that humanity's collective demands first surpassed the Earth's regenerative capacity around 1980. As of 2009 global demands on natural systems exceed their sustainable yield capacity by nearly 30 percent. This means we are meeting current demands in part by consuming the Earth's natural assets, setting the stage for an eventual Ponzi-type collapse when these assets are depleted.

As of mid-2009, nearly all the world's major aquifers were being overpumped. We have more irrigation water than before the overpumping began, in true Ponzi fashion. We get the feeling that we're doing very well in agriculture—but the reality is that an estimated 400 million people are today being fed by overpumping, a process that is by definition short-term. With aquifers being depleted, this water-based food bubble is about to burst.

A similar situation exists with the melting of mountain glaciers. When glaciers first start to melt, flows in the rivers and the irrigation canals they feed are larger than before the melting started. But after a point, as smaller glaciers disappear and larger ones shrink, the amount of ice melt declines and the river flow diminishes. Thus we have two water-based Ponzi schemes running in parallel in agriculture.

And there are more such schemes. As human and livestock populations grow more or less apace, the rising demand for forage eventually exceeds the sustainable yield of grasslands. As a result, the grass deteriorates, leaving the land bare, allowing it to turn to desert. In this Ponzi scheme, herders are forced to rely on food aid or they migrate to cities.

Three-fourths of oceanic fisheries are now being fished at or beyond capacity or are recovering from overexploitation. If we continue with business as usual, many of these fisheries will collapse. Overfishing, simply defined, means we are taking fish from the oceans faster than they can reproduce. The cod fishery off the coast of Newfoundland in Canada is a prime example of what can happen. Long one of the world's most productive fisheries, it collapsed in the early 1990s and may never recover.

Paul Hawken, author of *Blessed Unrest*, puts it well: "At present we are stealing the future, selling it in the present, and calling it gross domestic product. We can just as easily have an economy that is based on healing the future instead of stealing it. We can either create assets for the future or take the assets of the future. One is called restoration and the other exploitation." The larger question is: If we continue with business as usual, with overpumping, overgrazing, overplowing, overfishing, and overloading the atmosphere with carbon dioxide, how long will it be before the Ponzi economy unravels and collapses? No one knows. Our industrial civilization has not been here before.

Unlike Bernard Madoff's Ponzi scheme, which was set up with the knowledge that it would eventually fall apart, our global Ponzi economy was not intended to collapse. It is on a collision path because of market forces, perverse incentives, and poorly chosen measures of progress.

In addition to consuming our asset base, we have devised some clever techniques for leaving costs off the books—much like the disgraced and bankrupt Texas-based energy company Enron did some years ago. For example, when we use electricity from a coal-fired power plant we get a monthly bill from the local utility. It includes the cost of mining coal, transporting it to the power plant, burning it, generating the electricity, and delivering electricity to our homes. It does not, however, include any costs of the climate change caused by burning coal. That bill will come later—and it will likely be delivered to our children. Unfortunately for them, their bill for our coal use will be even larger than ours.

When Sir Nicholas Stern, former chief economist at the World Bank, released his groundbreaking 2006 study on the future costs of climate change, he talked about a massive market failure. He was referring to the failure of the market to incorporate the costs of climate change in the price of fossil fuels. According to Stern, the costs are measured in the trillions of dollars. The difference between the market prices for fossil fuels and an honest price that also incorporates their environmental costs to society is huge.

As economic decision makers we all depend on the market for information to guide us, but the market is giving us incomplete information, and as a result we are making bad decisions. One of the best examples of this can be seen in the United States, where the gasoline pump price was around $3 per gallon in mid-2009. This reflects only the cost of finding the oil, pumping it to the surface, refining it into gasoline, and delivering the gas to service stations. It overlooks the costs of climate change as well as the costs of tax subsidies to the oil industry, the burgeoning military costs of protecting access to oil in the politically unstable Middle East, and the health-care costs of treating respiratory illnesses caused by breathing polluted air. These indirect costs now total some $12 per gallon. In reality, burning gasoline is very costly, but the market tells us it is cheap.

The market also does not respect the carrying capacity of natural systems. For example, if a fishery is being continuously overfished, the catch eventually will begin to shrink and prices will rise, encouraging even more investment in fishing trawlers. The inevitable result is a precipitous decline in the catch and the collapse of the fishery.

Today we need a realistic view about the relationship between the economy and the environment. We also need, more than ever before, political leaders who can see the big picture. And since the principal advisors to government are economists, we need either economists who can think like ecologists or more ecological advisors. Otherwise, market behavior—including its failure to include the indirect costs of goods and services, to value nature's services, and to respect sustainable-yield thresholds—will cause the destruction of the economy's natural support systems, and our global Ponzi scheme will fall apart. ·

COAL

The Greatest Threat to Civilization

JAMES HANSEN

COAL IS THE *single greatest threat to civilization and all life on our planet. The climate is nearing tipping points. Changes are beginning to appear and there is a potential for explosive changes, effects that would be irreversible, if we do not rapidly slow fossil-fuel emissions over the next few decades. A moratorium on coal-fired power plants is by far the most important action that needs to be pursued.*

❖

In 2008, I wrote to former British prime minister Gordon Brown asking him to place a moratorium on new coal-fired power plants in Britain. I have asked the same of Angela Merkel, Barack Obama, Kevin Rudd, and other national leaders. The reason is this—coal is the single greatest threat to civilization and all life on our planet.

The climate is nearing tipping points. Changes are beginning to appear and there is a potential for explosive changes, effects that would be irreversible, if we do not rapidly slow fossil-fuel emissions over the next few decades. As Arctic sea ice melts, the darker ocean absorbs more sunlight and speeds melting. As the tundra melts, methane, a strong greenhouse gas, is released, causing more warming. As species are exterminated by shifting climate zones, ecosystems can collapse, destroying more species.

The public, buffeted by weather fluctuations and economic turmoil, has little time to analyze decadal changes. How can people be expected to evaluate and filter out advice emanating from those pushing special interests? How can people distinguish between top-notch science and pseudoscience?

Those who lead us have no excuse—they are elected to guide, to protect the public and its best interests. They have at their disposal the best scientific organizations in the world, such as the Royal Society and the U.S. National Academy of Sciences. Only in the past few years did the science crystallize, revealing the urgency. Our planet is in peril. If we do not change course, we'll hand our children a situation that is out of their control. One ecological collapse will lead to another, in amplifying feedbacks.

The amount of carbon dioxide in the air has already risen to a dangerous level. The pre-industrial carbon dioxide amount was 280 parts per million (ppm). Humans, by burning coal, oil, and gas, have increased this to 390 ppm; it continues to grow by about 2 ppm per year. Earth, with its four-kilometer-deep oceans, responds only slowly to changes of carbon dioxide. So the climate will continue to change, even if we make maximum effort to slow the growth of carbon dioxide emissions. Arctic sea ice will melt away in the summer season within the next few decades. Mountain glaciers, providing fresh water for rivers that supply hundreds of millions of people, will disappear—practically all of the glaciers could be gone within fifty years—if carbon dioxide continues to increase at current rates. Coral reefs, harboring a quarter of ocean species, are threatened.

The greatest danger hanging over our children and grandchildren is initiation of changes that will be irreversible on any timescale that humans can imagine. If coastal ice shelves buttressing the West Antarctic Ice Sheet continue to disintegrate, the sheet could disgorge into the ocean, raising sea levels by several meters in a century. Such rates of sea level change have occurred many times in Earth's history in response to global warming rates no higher than those of the past thirty years. Almost half of the world's great cities are located on coastlines.

The most threatening change, from my perspective, is extermination of species. Several times in Earth's history, rapid global warming occurred, apparently spurred by amplifying feedbacks. In each case, more than half of plant and animal species became extinct. New species came into being over tens and hundreds of thousands of years. But these are timescales and generations that we cannot imagine. If we drive our fellow species to extinction, we will leave a far more desolate planet for our descendants than the world we inherited from our elders.

Clearly, if we burn all fossil fuels, we will destroy the planet we know. Carbon dioxide would increase to 500 ppm or more. We would set the planet on a course to the ice-free state, with sea level 75 meters higher.

Climatic disasters would occur continually. The tragedy of the situation, if we do not wake up in time, is that the changes that must be made to stabilize the atmosphere and climate make sense for other reasons. They would produce a healthier atmosphere, improved agricultural productivity, clean water, and an ocean providing fish that are safe to eat.

Fossil fuel reservoirs will dictate the actions needed to solve the problem. Oil, of which half the readily accessible reserves have already been burned, is used in vehicles, so it's impractical to capture the carbon dioxide. This is likely to drive carbon dioxide levels to at least 400 ppm. But if we cut off the largest source of carbon dioxide—coal—it will be practical to bring carbon dioxide back to 350 ppm, lower still if we improve agricultural and forestry practices, increasing carbon storage in trees and soil.

Coal is not only the largest fossil fuel reservoir of carbon dioxide, it is the dirtiest fuel. Coal is polluting the world's oceans and streams with mercury, arsenic, and other dangerous chemicals. The dirtiest trick that governments play on their citizens is the pretense that they are working on "clean coal" or that they will build power plants that are "capture-ready" in case technology is ever developed to capture all pollutants.

The trains carrying coal to power plants are death trains. Coal-fired power plants are factories of death. They need to be shut down. The moratorium on any coal-burning power plants without carbon capture and sequestration (CCS) must begin in the West, which is responsible for three-quarters of climate change (via 75 percent of the present atmospheric carbon dioxide excess, above the pre-industrial level), despite large carbon dioxide emissions in developing countries.

The moratorium must extend to developing countries within a decade, but that will not happen unless developed countries fulfill their moral obligation to lead this moratorium. If Britain should initiate this moratorium, there is a strong possibility of positive feedback, a domino effect, with Germany, Europe, and the United States following, and then, probably with technical assistance, developing countries.

A spreading moratorium on construction of dirty (no CCS) coal plants is the sine qua non for stabilizing climate and preserving creation. It would need to be followed by phase-out of existing dirty coal plants in the next few decades, but would that be so difficult? Consider the other benefits: cleanup of local pollution, conditions in China and India now that greatly damage human health and agriculture, and present global export of pollution, including mercury that is accumulating in fish stocks throughout the oceans.

There are long lists of things that people can do to help mitigate climate change. But a moratorium on coal-fired power plants is by far the most important action that needs to be pursued. It should be the rallying issue for young people. The future of the planet in their lifetime is at stake. This is not an issue for only Bangladesh and the island nations, but for all humanity and other life on the planet.

THE VIEW FROM OIL'S PEAK

RICHARD HEINBERG

"PEAK OIL"—WHEN PETROLEUM extraction globally reaches its maximum and begins an inevitable decline—may be near, and the consequences are likely to be devastating to societies accustomed to abundant, inexpensive fossil fuels. Petroleum is the world's most important energy resource. There is no ready substitute, and decades will be required to wean societies from it. Peak oil could therefore pose the greatest economic challenge since the dawn of the Industrial Revolution.

❖

During the past decade a growing chorus of energy analysts has warned of the approach of "peak oil," when the global rate of petroleum extraction will reach its maximum and begin its inevitable decline. While there is some dispute among experts as to *when* this will occur, there is none as to *whether*. The global peak is merely the cumulative result of production peaks in individual oil fields and in oil-producing nations. The most important national peak occurred in the United States in 1970. At that time America produced 9.5 million barrels per day (mbd) of oil; the current figure is less than 6 mbd. While at one time the United States was the world's top oil-exporting nation, it is today the world's top importer.

The U.S. example helps in evaluating the prospects for delaying the global peak. After 1970, exploration efforts succeeded in identifying two enormous new American oil provinces—the North Slope of Alaska and the Gulf of Mexico. Meanwhile biofuels (principally ethanol) began to supplement crude. Also, improvements in oil recovery technology helped to increase the proportion of the oil in existing fields able to be extracted. These are the strategies (exploration, substitution, and technological improvements) that the energy industry is relying on either to delay the global production peak or to mitigate its impact. In the United States, each of these strategies made a difference—but not enough to reverse, for more than a few years now and then, a forty-year trend of declining production. The situation for the world as a whole is likely to be similar.

How near is the global peak? Today most oil-producing nations are seeing reduced output. In some instances, these declines are occurring because of lack of investment in exploration and production, or domestic political problems. But in most instances the decline results from factors of geology: While older oil fields continue to yield crude, beyond a certain point it becomes impossible to maintain maximum flow rates. Meanwhile, global rates of discovery of new oil fields have been declining since 1964.

These two trends—a growing preponderance of past-peak producers and a declining success rate for exploration—suggest that the world peak may be near. The consequences of peak oil are likely to be devastating. Petroleum is the world's most important energy resource. There is no ready substitute, and decades will be required to wean societies from it. Peak oil could therefore pose the greatest economic challenge since the dawn of the Industrial Revolution. For policy makers, five questions seem paramount:

1. How are the forecasts holding up?

While warnings about the end of oil were voiced in the 1920s and even earlier, the scientific study of petroleum depletion began with the work of geophysicist M. King Hubbert, who in 1956 forecast that U.S. production would peak within a few years of 1970 (in fact, that was the exact peak year), and who went on to predict that world production would peak close to the year 2000.

Shortly after Hubbert's death in 1989, other scientists issued their own forecasts for the global peak. Foremost among these were petroleum geologists Colin J. Campbell and Jean Laherrère, whose article "The End of Cheap Oil," published in *Scientific American* in March 1998, sparked the contemporary peak oil discussion. In the following decade, publications proliferated, including dozens of books, many peer-reviewed articles, websites, and film documentaries.

Most of the global peaking dates forecast by energy experts in the past few years have fallen within the decade from 2005 to 2015.[1] Running counter to these forecasts, IHS CERA, a prominent energy consulting firm, has issued reports foreseeing no peak before 2030.[2]

Are events unfolding in such a way as to support near-peak or the far-peak forecasts? According to the International Energy Agency, the past seven years have seen essentially flat production levels. These years have also seen extremely high oil prices, which should have provided a powerful incentive to increase production. The fact that actual crude oil production has not substantially increased during this period strongly suggests that the oil industry is near or has reached its capacity limits. It will be impossible to say with certainty that global oil production has peaked until several years after the fact. But the notion that it may already have reached its effective maximum must be taken seriously by policy makers.

2. What about other hydrocarbon energy sources?

If oil is becoming more scarce and less affordable, it would make sense to replace it with other energy sources, starting with those with similar characteristics—such as alternative hydrocarbons. There are very large amounts of total hydrocarbon resources; however, each is constrained by limits of various kinds. Bitumen (often called "oil sands" or "tar sands"), kerogen (sometimes referred to as "oil shale"), and shale oil (oil in low-porosity rocks that requires horizontal drilling and hydraulic fracturing for recovery) do not have the economic characteristics of regular crude oil, being more expensive to produce, delivering much lower energy return on investment, and entailing heavier environmental risks. Production from these sources may increase, but is not likely to offset declines in conventional crude over time.

Coal is commonly assumed to exist in nearly inexhaustible quantities. It could be used to produce large new amounts of electricity (with electric transport replacing oil-fueled cars, trucks, and trains), and it can be made into a liquid fuel. However, recent studies have shown that world coal reserves have been severely overestimated. Meanwhile, China's spectacular coal consumption growth virtually guarantees higher coal prices globally, making coal-to-liquids projects impractical.

Natural gas is often touted as a potential replacement for both oil and coal. However, conventional gas production in the United States is in decline. Unconventional gas production via hydraulic fracturing ("fracking") is increasing supplies over the short term, but this new production method is expensive and entails serious environmental risks; also, fracked gas wells deplete quickly, necessitating very high drilling rates.

Thus, while in principle there are several alternative hydrocarbon sources capable of substituting for conventional crude oil, all suffer from problems of quality and/or cost.

3. What might happen in the next decades absent policies to address peak oil?

The likely consequences of peak oil were analyzed at some length in the report, "Peaking of World Oil Production: Impacts, Mitigation, and Risk Management" (also known as the Hirsch Report), commissioned for the U.S. Department of Energy and published in 2005.[3] That report forecast "unprecedented" social, economic, and political impacts if efforts are not undertaken, at a "crash program" scale, and beginning at least a decade in advance of the peak, to reduce demand for oil and initiate the large-scale production of alternative fuels.

Clearly, the level of impact will depend partly on factors that can be influenced by policy. One factor that may *not* be susceptible to policy influence is the rapidity of the post-peak *rate of decline* in global oil production. The Hirsch Report simply assumed a 2 percent per year decline. In the first few years after peak, the actual decline may be smaller. That rate may increase as declines from existing fields accumulate and accelerate.

However, for some nations the situation may be much worse, since available oil export capacity will almost certainly contract faster than total oil production. Every oil-exporting nation also consumes oil, and domestic demand is typically satisfied before oil is exported. Domestic oil demand is growing in most oil-producing nations; thus the net amount available for export is declining even in some countries with steady overall production. Nations that are major oil importers, such as the United States, China, and many European nations, will feel strongly the effects of sharp declines in the amount of oil available on the export market.

High prices and actual shortages will dramatically impact national economies in several ways. The global transport system is almost entirely dependent on oil—not just private passenger automobiles, but trucks, ships, diesel locomotives, and the entire passenger and freight airline industry. High fuel prices will thus affect entire economies as travel becomes more expensive and manufacturers and retailers are forced to absorb higher transport costs.

Conventional industrial agriculture is also overwhelmingly dependent on oil, as modern farm machinery runs on petroleum products and oil is needed for the transport of farm inputs and outputs. Oil also provides the feedstock for making pesticides. According to one study, approximately seven calories of fossil fuel energy are needed to produce each delivered calorie of food energy in modern industrial food systems.[4] With the global proliferation of the industrial-chemical agriculture system, the products of that system are now also traded globally, enabling regions to host human populations larger than local resources alone could support. Those systems of global distribution and trade also rely on oil. Within the United States, the mean distance for food transport is now estimated at 1,546 miles.[5] High fuel prices and fuel shortages therefore translate to increasing food prices and potential food shortages.

A small but crucial portion of oil consumed globally goes into the making of plastics and chemicals. Some of the more common petrochemical building blocks of our industrial world are ethylene, propylene, and butadiene. Further processing of just these three chemicals produces products as common and diverse as disinfectants, solvents, antifreezes, coolants, lubricants, heat transfer fluids, and of course plastics, which are used in everything from building materials to packaging, clothing, and toys. Future oil supply problems will affect the entire chain of industrial products that incorporates petrochemicals.

Economic impacts to transport, trade, manufacturing, and agriculture will in turn lead to internal social tensions within importing countries. In exporting countries the increasing value of remaining oil reserves will exacerbate rivalries between political factions vying to control this source of wealth. Increased competition between consuming nations for control of export flows, and between importing nations and exporters over contracts and pipelines, may lead to international conflict. None of these effects is likely to be transitory. The crisis of peak oil will not be solved in months, or even years. Decades will be required to reengineer modern economies to function with a perpetually declining supply of oil.

4. How is the world responding?

In 1998, policy makers had virtually no awareness of peak oil as an issue. Now there are peak oil groups within the U.S. Congress and the British Parliament, and individual members of government in many other countries are keenly aware of the situation. Government reports have been issued in several nations.[6] Some cities have undertaken assessments of petroleum supply vulnerabilities and begun efforts to reduce their exposure.[7] A few nongovernmental organizations (NGOs) have been formed for the purpose of alerting government at all levels to the problem and helping develop sensible policy responses—notably, the Association for the Study of Peak Oil and Gas (ASPO) and the Post Carbon Institute. And grassroots efforts in several countries have organized "Transition Initiatives" wherein citizens participate in the development of local strategies to deal with the likely consequences of peak oil.

Unfortunately, this response is woefully insufficient given the scale of the challenge. Moreover, policies that *are* being undertaken are often ineffectual. Efforts to develop renewable sources of electricity are necessary to deal with climate change; however, they will do little to address the peak oil crisis, since very little of the transport sector currently relies on electricity that could be supplied from solar, wind, or other new electricity sources. Biofuels are the subject of increasing controversy having to do with ecological problems, the displacement of food production, and low energy efficiency; even in the best instance, they are unlikely to offset more than a small percentage of current oil consumption.

5. What would be an effective response?

One way to avert or ameliorate the impacts of peak oil would be to implement a global agreement to proactively reduce the use of oil (effectively, a reduction in *demand*) ahead of actual scarcity. Setting a bold but realistic mandatory target for demand restraint would reduce price volatility, aid with preparation and planning, and reduce international competition for remaining supplies. A proposal along these lines was put forward by physicist Albert Bartlett in 1986, and a similar one by petroleum geologist Colin Campbell in 1998; Campbell's proposal was the subject of the book *The Oil Depletion Protocol: A Plan to Avert Oil Wars, Terrorism and Economic Collapse.*[8] In order to enlist public support for such efforts, governments would need to devote significant resources to education campaigns. In addition, planning and public investment would be needed in transportation, agriculture, and chemicals-materials industries. For each of these there are two main strategic pathways.

Transportation
- Design communities to reduce the need for transportation (localize production and distribution of goods including food, while designing or redesigning urban areas for density and diversity);

- Promote alternatives to the private automobile and to air- and truck-based freight transport (by broadening public transport options, creating incentives for use of public transportation, and creating disincentives for automobile use). First priority should go to electrified transport options, as these are most efficient, then to alternative-fueled transport options, and finally to more-efficient petroleum-fueled transport options.

Agriculture

- Maximize local production of food in order to reduce the vulnerability implied by a fossil fuel–based food delivery system;
- Promote forms of agriculture that rely on fewer fossil fuel inputs.

Materials and Chemicals

- Identify alternative materials from renewable sources to replace petrochemical–based materials;
- Devise ways to reduce the amount of materials consumed.

Oil depletion presents a unique set of vulnerabilities and risks. If policy makers fail to understand these, nations will be mired in both internal economic turmoil and external conflict caused by fuel shortages. Policy makers may assume that, in addressing the dilemma of global climate change via carbon caps and trades, they would also be doing what is needed to deal with the problem of dependence on depleting petroleum. This could be a dangerously misleading assumption.

Fossil fuels have delivered enormous economic benefits to modern societies, but we are now becoming aware of the burgeoning costs of our dependence on these fuels. Humanity's central task for the coming decades must be the undoing of its dependence on oil, coal, and natural gas in order to deal with the twin crises of resource depletion and climate change. It is surely fair to say that fossil fuel dependency constitutes a systemic problem of a kind and scale that no society has ever had to address before. If we are to deal with this challenge successfully, we must engage in systemic thinking that leads to sustained, bold action.

ENERGY RETURN ON INVESTMENT

CHARLES A. S. HALL

ENERGY RETURN ON *Investment (EROI) is the ratio of energy returned from an energy-gathering activity compared to the energy invested in that process. While EROI by itself is not enough to judge the virtues or vices of particular fuels or energy sources, it is a crucial component for such assessments because it indicates whether a fuel is a net energy gainer or loser (and to what extent). EROI studies for most energy resources show a decline, indicating that depletion has been more important than technological improvements over time.*

❖

Energy return on investment (EROI)[1], or sometimes "energy return on energy investment" (EROEI), is the ratio of energy returned from an energy-gathering activity compared to the energy invested in that process. (The word "investment" usually means energy investment, but sometimes net energy analysis also includes financial, environmental, and/or other kinds of investments.) The term EROI has been around since at least 1970, but it gained relatively little traction until the last five or ten years. Now there is an explosion of interest as peak oil and the general economic effects of increasingly constrained energy supplies are becoming obvious to investigators from many fields. Many observers feel that the financial crises we have been experiencing since 2008 are a direct effect of the end of oil production growth (of all liquid fuels if considered on an energy basis) and of the general decline in EROI for most energy sources.

While EROI by itself is not enough to judge the virtues or vices of particular fuels or energy sources, it is an extremely important component for such assessments. Most importantly, it can indicate if a fuel is a net energy gainer or loser—and to what extent. It also offers the possibility of looking into the future in a way that markets are unable to do: EROI advocates believe that, in time, market prices must approximately reflect comprehensive EROIs (at least if appropriate corrections for energy quality are made and financial subsidies for energy and fuel production are removed).[2]

THE IMPORTANCE OF EROI

Many prominent earlier researchers and thinkers (including sociologists Leslie White and Fred Cottrell, ecologist Howard Odum, and economist Nicolas Georgescu Roegen) have emphasized the importance of net energy and energy surplus as a determinant of human culture. Farmers and other food producers must have an energy profit for there to be specialists, military campaigns, and cities, and substantially more for there to be art, culture, and other amenities. Net energy analysis is simply a way of examining how much energy is left over from an energy-gaining process after correcting for how much of that energy— or its equivalent from some other source—is required to generate a unit of the energy in question.

The importance of EROI is far more than simply whether it is positive or negative. Several of the participants in the current debate about corn-derived ethanol believe that corn-based ethanol has an EROI of less than 1:1, while others argue that ethanol from corn shows a clear energy surplus, with from 1.2 to 1.6 units of energy delivered for each unit invested. But this argument misses a very important issue. Think of a society dependent upon one resource: oil. If the EROI for this oil was 1.1:1 then one could pump the oil out of the ground and look at it…and that's it. It would be an energy loss to do anything else with it. If it were 1.2:1 you could refine it into diesel fuel, and at 1.3:1 you could distribute it to where you want to use it. If you actually want to run a truck with it, you must have an EROI ratio of at least 3:1 (at the wellhead) to build and maintain the truck, as well as the necessary roads and bridges (including depreciation). If additionally you wanted to put something in the truck and deliver it, that would require an EROI of, say, 5:1.[3] Now say you wanted to include depreciation on the oil field worker, the refinery worker, the truck driver, and the farmer; you would need an EROI of 7:1 or 8:1. If their children were to be educated you would need perhaps 9:1 or 10:1, to have health care 12:1, to have arts in their lives maybe 14:1, and so on.

Obviously to have a modern civilization one needs not just surplus energy, but lots of it—and that requires either a high EROI or a massive source of moderate-EROI fuels. If these are not available, the remaining low-EROI energy will be prioritized for growing food and supporting families.

If the energy and hence economic pie is no longer getting larger—indeed, if because of geological constraints it can no longer get larger—how will we slice it? This may force some ugly debates back into the public vision. If EROI continues to decline then it will cut increasingly into discretionary spending (the engine for economic growth) and we will need to ask some very hard questions about how we should spend our money.

A problem with substitutes to fossil fuels is that, of the alternatives currently available, none appear to have all the desirable traits of fossil fuels, especially liquids. These include sufficient energy density, easy transportability, relatively low environmental impact per net unit delivered to society, relatively high EROI, and availability on a scale that society presently demands. Thus it would seem that the United States and the rest of the world are likely facing a decline in both the quantity and EROI of its principal fuels. How we adjust to this will be a critical determinant of our future.

THE ECONOMIC COST OF ENERGY

In real economies, energy is essential for any process to occur; that is, for the production and transport of goods and services (even for the production of financial services). In the United States, that necessary energy comes from many sources: from imported and domestic sources of oil (about 40 percent), coal and natural gas (about 20 percent each), from hydropower and nuclear (about 5 percent each), and from a little renewable energy (mostly as firewood but increasingly from wind and solar).

It is possible to examine the ratio of the cost of energy (from all sources, weighed by their importance) relative to the benefits of using it to generate wealth. In 2007, roughly 9 percent of gross domestic product (GDP) was spent to purchase the energy used by the U.S. economy to produce the goods and services that comprised the GDP. Over recent decades that ratio has varied between 5 and 14 percent. The abrupt rise and subsequent decline in the proportion of the GDP spent for energy was seen during the "oil shocks" of the 1970s, in mid-2008, and again in 2011. Each of these increases in the price of oil relative to GDP had large impacts on discretionary spending—that is, on the amount of income that people can spend on what they want versus what they need. An increase in energy cost from 5 to 10 or even 14 percent of GDP would come mainly out of the 25 percent or so of the economy that usually goes to discretionary spending. Thus changes in the amount we spend on energy (much of which goes overseas) have very large impacts on the U.S. economy since most discretionary spending is domestic. This is why each significant increase in the price of oil (and of energy generally) has been associated with an economic recession, and it suggests that declining EROI will take an increasing economic toll in the future.[4]

DETERMINING EROI

EROI is calculated from the following simple equation, although the devil is in the details:

$$\text{EROI} = \frac{\text{Energy returned to society}}{\text{Energy required to get that energy}}$$

The numerator and denominator are necessarily assessed in the same units so that the ratio derived is dimensionless (e.g., 30:1). This example means that a particular process yields 30 joules per investment of 1 joule (or kilocalorie per kilocalorie, or barrels per barrel). EROI is usually applied at the point that the energy resource leaves the extraction or production facility (i.e., at the mine-mouth, wellhead, farm gate, etc.); we denote this more explicitly as EROI_{mm}. Another approach uses a simple, standardized energy output divided by the direct energy (energy used at the production site) and indirect energy (energy used to manufacture the machinery and products used at the production site) consumed to generate that output. This results in a measurement of standard EROI, EROI_{st}.

Determining the energy content of the numerator of the EROI equation is usually straightforward: Simply multiply the quantity of energy produced by the energy content per unit. Determining the energy content of the denominator is usually more difficult. The energy used directly (i.e., on site) might include, for example, the energy used to rotate the drilling bit when drilling for oil, the energy used to excavate when mining for coal, or the energy used to operate the farm tractor when harvesting corn for ethanol. One also should include the energy used indirectly—that is, the energy used to manufacture the drilling bit, the excavation equipment, the tractor, and so on. Companies generally do not keep track of their energy expenditures in terms of joules, only in dollars. However, it is possible to convert dollars spent to energy spent using either the mean price of fuel for direct energy or by using "energy intensities" for dollars spent in different parts of the economy (such as were calculated by a University of Illinois research group in the 1970s[5]).

Of course, the EROI that is needed to profitably undertake some economic activity, such as driving a truck, is far more than just what is needed to get the fuel out of the ground. Starting with the EROI for an energy source from the point of production ($EROI_{mm}$), we can then consider what might be needed to refine that energy source and deliver it to its point of use; we could also include the (prorated) energy required to make and maintain a vehicle and the roads it would drive on. This would give us EROI at the "point of use," or $EROI_{pou}$: the ratio of energy available at a point of use to the energy required for acquiring and delivering that energy.

EROI OF OBTAINING ENERGY THROUGH TRADE

An economy without enough domestic fuels of the type it needs must import the fuels and pay for them with some kind of surplus economic activity. Thus the economy's ability to purchase the required energy depends upon what it can generate to sell to the world, as well as upon the fuel required to grow or produce that material. The EROI for the imported fuel is the relation between the amount of fuel bought with a dollar relative to the amount of dollar profits gained by selling goods or services for export. The quantity of the goods or services that need to be exported to attain a barrel of oil depends upon the relative prices of the fuel versus the exported commodities.

In the 1980s, Boston University scholar Robert Kaufmann estimated the energy cost of generating a dollar's worth of major U.S. exports (e.g., wheat, commercial jetliners), and also the chemical energy found in one dollar's worth of imported oil.[6] The concept was that the EROI for imported oil depended upon what proportion of an imported dollar's worth of oil you needed to generate the money from overseas sales that you traded, in a net sense, for that oil. He concluded that, before the oil price increases of the 1970s, the EROI for imported oil was about 25:1 (very favorable for the United States); but this dropped to about 9:1 after the first oil price hike in 1973 and then to about 3:1 following the second oil price hike in 1979. The ratio returned to more favorable levels (from the perspective of the United States) from 1985 to about 2000 as the price of exported goods increased through inflation more rapidly than the price of oil. As oil prices increased again in the first decade of the twenty-first century—a period when the remaining conventional oil became concentrated in fewer and fewer countries, and future supply of conventional oil was of increasing concern—that ratio declined again to roughly 10:1. Estimating the EROI of obtaining energy through trade may be very useful in predicting economic vulnerability for specific countries in the near future.

To some degree we have managed to continue purchasing foreign oil through debt, which gives us a temporarily higher EROI. Were we to pay off this debt in the future, and if those who got the dollars wished to turn them into real goods and services (which seems a reasonable assumption), then we would have to take some substantial part of our remaining energy reserves out of the ground and convert it into fish, rice, beef, cars, and other products that those people would be able to buy from us.

THE HISTORY OF EROI

EROI has precedents in the concept of "net energy analysis" used by Leslie White, Kenneth Boulding, and especially Howard Odum.[7] Its origins were derived most explicitly in my 1970 doctoral dissertation on the energy costs and gains of migrating fish.[8] The concept was developed in various papers throughout the 1980s, and although its use lagged during society's "energy lull" from 1984 to 2005 it has since picked up significantly.[9] Similar but less explicit and focused ideas can be found in the newer field of "life cycle analysis," which is better developed in Europe than in the United States.

There have been questions about the degree to which we should use EROI versus more familiar measurements (e.g., financial return on financial investment in the oil business) to examine energy and other resource choices. In addition there have been criticisms that EROI has some severe flaws, such as that different studies give different answers to what appears to be the same question, that the boundaries of the analysis are controversial, that market solutions are always superior to "contrived" scientific studies, and that EROI too often is dependent upon monetary data for its results. Despite these real or imagined limitations, EROI is still a critical concept to understand when considering energy policy and the future prospects for modern civilization.

ALTERNATIVE ENERGY CHALLENGES

DAVID FRIDLEY

ALTERNATIVE ENERGY DEPENDS heavily on engineered equipment and infrastructure for capture or conversion. However, the full supply chain for alternative energy, from raw materials to manufacturing, is still very dependent on fossil fuel energy. The various obstacles to alternative energy—scalability, substitutability, intermittency, energy density, and others—compound the fundamental challenge of how to supplant a fossil fuel–based supply chain with one driven by alternative energy forms themselves.

❖

High oil prices, concerns over energy security, and the threat of climate change have all stimulated investment in the development of alternatives to conventional oil. "Alternative energy" generally falls into two categories:

- Substitutes for existing petroleum liquids (ethanol, biodiesel, biobutanol, dimethyl ether, coal-to-liquids, tar sands, oil shale), both from biomass and fossil feedstocks; and
- Alternatives for generating and storing electric power (wind, solar photovoltaics, solar thermal, tidal, biomass, fuel cells, batteries).

The technology pathways to these alternatives vary widely, from distillation and gasification to bioreactors of algae and high-tech manufacturing of photon-absorbing silicon panels. Many are considered "green" or "clean" although some, such as coal-to-liquids and tar sands, are "dirtier" than the petroleum they are replacing. Others, such as biofuels, have concomitant environmental impacts that offset potential carbon savings.

Unlike conventional fossil fuels, where nature provided energy over millions of years to convert biomass into energy-dense solids, liquids, and gases—requiring only extraction and transportation technology for us to mobilize them—alternative energy depends heavily on engineered equipment and infrastructure for capture or conversion, essentially making it a high-tech manufacturing process. However, the full supply chain for alternative energy, from raw materials to manufacturing, is still very dependent on fossil fuel energy for mining, transport, and materials production. Alternative energy faces the challenge of how to supplant a fossil fuel–based supply chain with one driven by alternative energy forms themselves in order to break their reliance on a fossil fuel foundation.

The public discussion about alternative energy is often reduced to an assessment of its monetary costs versus those of traditional fossil fuels, often in comparison to their carbon footprints. This kind of reductionism to a simple monetary metric obscures the complex issues surrounding the potential viability, scalability, feasibility, and suitability of pursuing specific alternative technology paths. Although money is necessary to develop alternative energy, money is simply a token for mobilizing a range of resources used to produce energy. At the level of physical requirements, assessing the potential for alternative energy development becomes much more complex since it involves issues of end-use energy requirements, resource use trade-offs (including water and land), and material scarcity.

Similarly, it is often assumed that alternative energy will seamlessly substitute for the oil, gas, or coal it is designed to supplant—but this is rarely the case. Integrating alternatives into our current energy system will require enormous investment in both new equipment and infrastructure—along with the resources required for their manufacture—at a time when capital to make such investments has become harder to secure. This raises the question of the suitability of moving toward an alternative energy future with an assumption that the structure of our current large-scale, centralized energy system should be maintained. Since alternative energy resources vary greatly by location, it may be necessary to consider different forms of energy for different localities.

Assessing the promise of alternative energy is complex and multi-faceted; the discussion is complicated by political biases, ignorance of basic science, and a lack of appreciation of the magnitude of the problem facing societies accustomed to inexpensive fossil energy as the era of abundance concludes. While not a comprehensive listing, the key challenges of alternative energy include:

SCALABILITY AND TIMING

For the promise of an alternative energy source to be achieved, it must be supplied in the time frame needed, in the volume needed, and at a reasonable cost. Many alternatives have been successfully demonstrated at the small scale (algae-based diesel, cellulosic ethanol, biobutanol, thin-film solar), but demonstration scale does not provide an indication of the potential for large-scale production. Similarly, because alternative energy relies on engineering, manufacturing, and construction of equipment and manufacturing processes for its production, output grows in a step-wise function only as new capacity comes online, which in turn is reliant on timely procurement of the input energy and other required input materials. This difference between "production" of alternative energy and "extraction" of fossil fuels can result in marked constraints on the ability to increase the production of an alternative energy source as it is needed.

COMMERCIALIZATION

Closely related to the issue of scalability and timing is commercialization, or the question of how far away a proposed alternative energy source stands from being fully commercialized. Often, newspaper reports of a scientific laboratory breakthrough are accompanied by suggestions that such a breakthrough represents a possible "solution" to our energy challenges. In reality, the average time frame between laboratory demonstration of feasibility and large-scale commercialization is from twenty to twenty-five years. Processes need to be perfected and optimized, patents developed, demonstration tests performed, pilot plants built and evaluated, environmental impacts assessed, and engineering, design, siting, financing, economic, and other studies undertaken.

SUBSTITUTABILITY

Ideally, an alternative energy form would integrate directly into the current energy system as a "drop-in" substitute for an existing form without requiring further infrastructure changes. This is rarely the case, and the lack of substitutability is particularly pronounced in the case of electric vehicles. Although it is possible to generate the needed electricity from wind or solar power, the prerequisites to achieving this are extensive. Electric car proliferation at a meaningful scale would require extensive infrastructure changes including retooling factories to produce the vehicles, developing a large-scale battery industry and recharging facilities, building a maintenance and spare parts industry, integrating "smart grid" monitoring and control software and equipment, and of course, constructing additional generation and transmission capacity. All of this is costly.

The development of wind and solar power electricity also requires additional infrastructure; wind and solar electricity must be generated where the best resources exist, which is often far from population centers. Thus extensive investment in transmission infrastructure to bring it to consumption centers is required. Today, ethanol can be blended with gasoline and used directly, but its propensity to absorb water and its high oxygen content make it unsuitable for transport in existing pipeline systems,[1] and an alternative pipeline system to enable its widespread use would be materially and financially intensive. While alternative energy forms may provide the same energy services as another form, they rarely substitute directly, and these additional material costs need to be considered.

MATERIAL INPUT REQUIREMENTS

The key input to an alternative energy process is not money, but resources and energy; the type and volume of the resources and energy needed may in turn limit the scalability and affect the cost and feasibility of an alternative. This is particularly notable in processes that rely on advanced technologies manufactured with rare earth elements. Fuel cells, for example, require platinum, palladium, and rare earth elements. Solar photovoltaic technology requires gallium, and in some forms, indium. Advanced batteries rely on lithium. Even technology designed to save energy, such as LED or organic LED (OLED) lighting, requires the rare earths indium and gallium. Expressing the costs of alternative energy only in monetary terms obscures potential limits from the resource and energy inputs required. Successful deployment of a range of new energy technologies (and some nonenergy advanced technologies) would substantially raise demand for a range of metals beyond the level of world production today.

Alternative energy production is reliant not only on a range of resource inputs, but also on fossil fuels for the mining of raw materials, transport, manufacturing, construction, maintenance, and decommissioning. Currently, no alternative energy exists without fossil fuel inputs, and no alternative energy process can reproduce itself—that is, manufacture the equipment needed for its own production—without the use of fossil fuels. In this regard, alternative energy serves as a supplement to the fossil fuel base, and its input requirements may constrain its development in cases of either material or energy scarcity.

INTERMITTENCY

Modern societies expect that electrons will flow when a switch is flipped, that gas will flow when a knob is turned, and that liquid fuel will flow when the pump handle is squeezed. This system of continuous supply is possible because of our exploitation of large stores of fossil fuels, which are the result of millions of years of intermittent sunlight concentrated into a continuously extractable source of energy. Alternative energies such as solar or wind power produce only intermittently as the Sun shines or the wind blows, and even biomass-based fuels depend on seasonal harvests of crops. Integrating these energy forms into our current system creates challenges of balancing availability and demand, and it remains doubtful that these intermittent energy forms can provide a majority of our future energy needs in the same way that we expect energy to be available today.

The key to evening out the impact of intermittency is storage; that is, developing technologies and approaches that can store energy generated during periods of good wind and sun for use at other times. Many approaches have been proposed and tested, including compressed air storage, batteries, and the use of molten salts in solar thermal plants. The major drawbacks of all these approaches include the losses involved in energy storage and release, and the limited energy density that these storage technologies can achieve.

ENERGY DENSITY

Energy density refers to the amount of energy that is contained in a unit of an energy form. It can be expressed in the amount of energy per unit of mass (weight), or in the amount of energy per unit of volume. Energy density has greatly influenced our choice of fuels. The conversion to the use of coal in the seventeenth and eighteenth centuries was welcomed because coal provided twice as much energy as wood for the same weight of material. Similarly, the shift from coal to petroleum-powered ships in the early twentieth century was driven by the fact that petroleum possesses nearly twice the energy density of coal, allowing ships to go farther without having to stop for refueling. Even when used in a motor vehicle's inefficient internal combustion engine, a kilogram of highly energy-dense gasoline—about six cups—allows us to move 3,000 pounds of metal roughly 11 miles.

The consequence of low energy density is that larger amounts of material or resources are needed to provide the same amount of energy as a denser material or fuel. Many alternative energies and storage technologies are characterized by low energy densities, and their deployment will result in higher levels of resource consumption. Lithium ion batteries—the focus of current research for electric vehicles—can contain only 0.5 megajoules per kilogram (MJ/kg) of battery compared to 46 MJ/kg for gasoline. Advances in battery technology are being announced regularly, but they all come up against the theoretical limit of energy density in batteries of only 3 MJ/kg.

ENERGY RETURN ON INVESTMENT

The complexity of our economy and society is a function of the amount of net energy we have available. "Net energy" is, simply, the amount of energy remaining after we consume energy to produce energy. Consuming energy to produce energy is unavoidable, but only that which is not consumed to produce energy is available to sustain our industrial, transport, residential, commercial, agricultural, and military activities. The ratio of the amount of energy we put into energy production and the amount of energy we produce is called "energy return on investment" (EROI). EROI can be very high (e.g. 100:1, or 100 units of energy produced for every one unit used to produce it—an "energy source"), or low (0.8:1, or only 0.8 units of energy produced for every one unit used in production—an "energy sink"). Society requires energy sources, not energy sinks, and the magnitude of EROI for an energy source is a key indicator of its contribution to maintenance of social and economic complexity.

Net energy availability has varied tremendously over time and in different societies. In the last advanced societies that relied only on solar power (sun, water power, biomass, and the animals that depended on biomass) in the seventeenth and early eighteenth centuries, the amount of net energy available was low and dependent largely on the food surpluses provided by farmers. At that time, only 10 to 15 percent of the population was *not* involved in energy production. As extraction of coal, oil, and natural gas increased in the nineteenth and twentieth centuries, society was increasingly able to substitute the energy from fossil fuels for manual or animal labor, thereby freeing an even larger proportion of society from direct involvement in energy production. In 1870, 70 percent of the U.S. population was farmers; today the figure is less than 2 percent, and every aspect of agricultural production now relies heavily on petroleum or natural gas. The same is true in other energy sectors: Currently, less than 0.5 percent of the U.S. labor force (about 710,000 people) is directly involved in coal mining, oil and gas extraction, petroleum refining, pipeline transport, and power generation, transmission, and distribution.

The challenge of a transition to alternative energy, then, is whether such energy surpluses can be sustained, and thus whether the type of social and economic specialization we enjoy today can be maintained. Indeed, one study estimates that the minimum EROI for the maintenance of industrial society is 5:1, suggesting that no more than 20 percent of social and economic resources can be dedicated to the production of energy without undermining the structure of industrial society.[2] In general, most alternative energy sources have low EROI values. A high EROI is not sufficient to ensure that the structure of modern society and economies can be maintained, but it is a prerequisite.

CONCLUSION

Alternative energy forms are crucial for a global transition away from fossil fuels, despite the myriad challenges of their development, scaling, and integration. In face of the peaking of global oil production—to be followed by peaks in natural gas and coal extraction—and of the need to reverse trajectory in carbon emissions, alternative energy sources will need to form the backbone of a future energy system.

That system, however, will not be a facsimile of the system we have today based on continuous uninterrupted supply growing to meet whatever demand is placed on it. As we move away from the energy bounty provided by fossil fuels, we will become increasingly reliant on tapping the current flow of energy from the Sun (wind, solar) and on new energy manufacturing processes that will require ever-larger consumption of resources (biofuels, other manufactured liquids, batteries). What kind of society we can build on this foundation is unclear, but it will most likely require us to pay more attention to controls on energy demand to accommodate the limitations of our future energy supply. Moreover, the modern focus on centralized production and distribution may be hard to maintain, since local conditions will become increasingly important in determining the feasibility of alternative energy production.

WHEN RISK ASSESSMENT IS RISKY

Predicting the Effects of Technology

DAVID EHRENFELD

EACH NEW TECHNOLOGY, *regardless of benefits, brings its own risks. In many complex situations where there are multiple questions with poorly constrained answers, it is folly to expect that we can use formal risk assessments to guide current actions. Risk assessment can be done properly if it does not pretend a false omniscience, and is considered in its full political, economic, social, biological, and moral context.*

❖

Technology has its own momentum. Propelled faster and faster by profits, politics, and the thrill of discovery, it has gone out of control, although this is a misleading way of putting it because it has never been under control. Each new technology, regardless of benefits, brings its own risks: extinctions, explosions, nuclear plant meltdowns, chemical toxicities, soil degradation, invasive species, climate change, resource depletion, damage to human communities. Who has not been made aware of some or all of these mortal threats? Everyone has experienced the effects of technology—bad and good. We know or sense that new technologies bring new risks, and this can produce an understandable sense of danger, a fear of an unknown future.

Throughout history, people have worried about what is going to happen next. The Greeks had oracles, notably the one at Apollo's temple in Delphi. The Delphic oracle had alleged successes—it also had noteworthy failures, such as the prediction of victory for the Persians in the fifth-century BCE war with the Greeks. But Apollo's priestesses, known as Pythias, uttered prophecies for more than one thousand years, so the oracle must have gotten some things right. Its last recorded message, for the Roman Emperor Julian the Apostate, who tried to restore Apollo's temple and oracle in 362 CE, was a chilling description of the end of Greek prophecy:

Tell ye the king: the carven hall is fallen in decay;
Apollo hath no chapel left, no prophesying bay,
No talking spring. The stream is dry that had so much to say.[1]

Of course, biblical prophecy has had an even longer run than the Greek oracles. The Hebrew prophets are still actively studied and invoked, although their writings do not seem suitable for predicting the specific effects of most current technologies. Now, because of the tremendous power of technology to change the world quickly and dramatically, we feel a new and very urgent need to predict the future accurately.

A very modern substitute for prophecy is the widespread belief that because technology can do so many extraordinary things it can predict its own side effects, letting us make effective plans to avoid future disasters. Technology is moving ahead whether we like it or not, and it is comforting to believe that we can figure out the problems we are likely to face, that we are the engineers of the hurtling freight train, not passengers locked in a windowless baggage car.

We don't rely on oracles; we now have scientific models that give us apparently reliable risk assessments and other quantitative probabilities of the occurrence of future events. The likely date of the extinction of an endangered species, the number of people who will be affected by an epidemic, the world's human population fifty years from now, the demand for electric power in 2030, the trajectory of the Dow Jones Industrial Average over the next three months—these are just a few of the kinds of predictions with specific numerical ranges. But as was the case with the Delphic Oracle, scientific prediction has had many failures. Twentieth-century predictions of the size of Canadian and New England cod populations based on models of sustainable yield, estimates made in the 1960s of U.S. electric consumption in the 1990s, most predictions of the robust growth of the economy made in early 2007, and reassuring risk assessments of the safety of injecting into the earth (then removing) millions of gallons of water and toxic chemicals to stimulate extraction of natural gas from the Marcellus shale formations in Pennsylvania and New York—all proved wrong, sometimes disastrously wrong.

Some of these forecasting failures have occurred because the data used to calculate the risk were preselected to give a politically or economically desired result—the books were cooked. Some forecasts fail because of statistical problems with predicting extremely infrequent outcomes (the

"base rate fallacy"). Other predictions, including those calculated hon-
estly, have been rendered worthless by the occurrence of rare, unpredictable happenings that lie outside the realm of our common experience.
Predictions can fall victim to a combination of these factors. And as financial trader Nassim Taleb has written, "it does not matter how frequently
[predictions succeed] if failure is too costly to bear."[2] It is not clear which
of the two systems, prophecy or scientific assessment, will work better
for many of the most critical predictions we have to make. We usually
assume that the scientific method will give us the correct answers if we
ask in the right way. Often it does, but frequently it doesn't. Can we tell
ahead of time which of our forecasts will be reasonable and which will
not? To date, there have been few systematic efforts to check whether
previous forecasts were right or wrong, no track record to help us evaluate
the reasonableness of new predictions. Climate scientists, who are subject
to enormous public and scientific scrutiny, frequently make reality checks
of their past predictions, but they are an exception. More often, we lack
a forecasting track record to help us evaluate new predictions. Without
quality control for our scientific predictions, we will make the same mis-
takes over and over again.

After the explosion of BP's Deepwater Horizon oil rig, which had
been drilling the Macondo well in the Gulf of Mexico on April 20, 2010,
we heard a good deal about risk assessment. Coast Guard Admiral Thad
W. Allen, incident commander for the oil spill remediation, properly
called for "a responsible assessment of the risks and benefits" of the various
damage control tactics that were being used. He wanted to know how
much oil they would remove and what their ultimate impact would be,
years later, on life in the Gulf.

Important decisions hinge on assessments like the ones that Admiral
Allen requested. Critical forecasts of this sort are made every day around
the world. Some will be right and some will be grotesquely wrong. Can
we find a general method for assessing the trustworthiness of a prediction
before we start using it to set policy? The answer is yes—and we can start
by taking a closer look at the process of risk assessment itself.

There are two kinds of risk assessment; they are quite different. The
first, common in structural engineering, evaluates completely defined
risks in systems that are visible and open to inspection. For example: At
what point will this bridge collapse as we increase the load? Is this building

likely to withstand an earthquake of magnitude 6.5? Risk assessments of this kind are proper and necessary, although like any scientific estimate they are subject to error.

The second type of risk assessment addresses complex systems with many interacting, tightly coupled subsystems, some of which are either poorly understood or beyond our power of inspection. These systems are subject to unexpected and unpredictable breakdowns that the Yale sociologist Charles Perrow called "normal accidents."[3] The Deepwater Horizon fell in this category: The critical events occurred nearly a mile beneath the ocean surface and were brought about or compounded by a confusion of conflicting command structures among BP, Halliburton, Transocean, and the Minerals Management Service.

The first major risk assessment of a complex system was WASH-1400, a report on nuclear reactor safety produced in 1975 for the Nuclear Regulatory Commission by a team headed by professor Norman Rasmussen at the Massachusetts Institute of Technology. The authors concluded that the risk of core meltdown was one in 20,000 per reactor per year, acceptably small. Around the same time the report was issued, the largest nuclear plant in the United States, in Browns Ferry, Alabama, was set on fire and temporarily put out of commission by a worker using a candle to check for air leaks. Core meltdown was narrowly averted. Had Dr. Rasmussen's team thought about including the effects of candles in their risk assessment? Of course not. Similarly, when DDT was first introduced, did anyone think to ask whether it would chemically mimic sex hormones, weakening the eggshells of ospreys and eagles? Of course not. The limitations of present knowledge can prevent us from asking all the risk-relevant questions. And if you can't think of all the important questions to ask, you can't make a useful risk analysis.

We have to try to predict the effects of the oil dispersants and other spill-control technologies on marine life and coastal marshes, but let's not kid ourselves that such predictions are more than informed guesses. We can't evaluate remedies if we don't even know every place where the dispersant-treated oil is going, or its effects on many thousands of species at different depths in the deep ocean, around reefs, and in intertidal zones. We don't really know the long-term effects of other proposed strategies such as the release of genetically engineered oil-eating bacteria, and the addition of huge quantities of nitrogen and phosphate compounds to stimulate the bacteria's growth. No lab experiment can be scaled up from a test tube to an ocean to give us complete answers.

In many complex situations where there are multiple questions with poorly constrained answers, it is folly to expect that we can use formal risk assessments to guide current actions. Fortunately, there is a way to determine those cases in which relying on risk analysis is inappropriate or dangerous. There are good criteria for rejecting some "scientific" predictions before they have a chance to do damage. The safety of using risk assessments decreases dramatically when: 1) the proposed strategies for preventing problems and minimizing damage have never been tested under real conditions; 2) when many interacting systems—especially ecological and human social systems—are involved; 3) when events that are beyond our inspection or even comprehension are likely to happen; 4) when the prediction extends unrealistically far into the future; and 5) when there are side effects of the proposed action that are already known to be serious. Not all of these five conditions need apply to invalidate a risk analysis; there is plenty of latitude for informed debate and judgment.

When confronted by an authoritative-seeming prediction, we should first ask whether it concerns a simple or complex system. This tells us how far we can extend our trust. For example, forecasts of eclipses to take place one hundred years from now are likely to be reliable. Astronomical calculations, appearances to the contrary, deal with relatively simple, well-described, purely physical systems. On the other hand, long-term weather forecasts (those extending beyond a few weeks) are often wrong. The weather is influenced by so many different variables on land, in the ocean, and in the atmosphere, and by so many random events, that detailed long-term forecasts are not believable unless they are so general as to be useless (it will be warmer in August than in November). Geologists Orrin Pilkey and Linda Pilkey-Jarvis have written in their book *Useless Arithmetic: Why Environmental Scientists Can't Predict the Future,*

> *If we wish to stay within the bounds of reality we must look to a more qualitative future, a future where there will be no certain answers to many of the important questions we have about the future of human interactions with the earth. ... No one knows in what order, for what duration, from what direction, or with what intensity the various events that affect a process will unfold. No one can ever know.*[4]

A spurious aura of certainty about the future has allowed us to be imposed upon by the proponents of technology at all costs. Accident follows accident. The damages, some of them irreversible, accumulate. But if we strip away the veneer of certainty and reject the pseudoscientific, self-serving calculations about the future—if we remember the unlucky nuclear-plant worker with his candle—we free ourselves to make responsible decisions about our destiny and the destiny of the natural world. We free ourselves to listen to the advice of conflicting experts, and to ask the hard questions that ought to be asked when they should be asked. We are less likely to be stampeded into making decisions that imperil our future as well as the future of myriad other species. We can rely more on adaptive management that is responsive to the situation as it exists now, without prejudice by long-term forecasts. And we can make contingency plans that assume that our risk assessments may be wrong.

Risk analyst Paul Slovic has noted: "Risk assessment is inherently subjective and represents a blending of science and judgment with important psychological, social, cultural, and political factors." He advocates "introducing more public participation into both risk assessment and risk decision making in order to make the decision process more democratic, improve the relevance and quality of technical analysis, and increase the legitimacy and public acceptance of the resulting decisions."[5] If such an atmosphere of sanity had prevailed during the first submission of corporate risk assessments that claimed it was safe for Transocean's mobile offshore unit to sink an oil well a mile deep in the Gulf, drilling for the ill-starred Macondo well would probably have been stopped before it began. Then there would have been no oiled pelicans gasping for breath, drowned or burned sea turtles, ruined coastal economies, and desperate fishermen seeing the end of their life's work and the destruction of the cultural legacies of their parents, grandparents, and great-grandparents.

Knowing the limitations of prediction and risk analysis in the complex situations posed by modern technology does not have to paralyze our ability to plan and act. It allows us to use our science wisely and carefully, and to introduce into our decisions the values of concern for people and concern for nature that alone can make our technology safe and lasting. Risk assessment can be done properly if it does not pretend a false omniscience and is considered in its full political, economic, social, biological, and moral context. Above all, we can never let the dispassionate, academic nature of risk assessments make us forget that bad predictions can cause lasting misery for us and for the Earth we inhabit.

MALEVOLENT AND MALIGNANT THREATS

R. JAMES WOOLSEY

MALEVOLENT THREATS TO *national security are the product of human choice, such as terrorists choosing to attack our electricity grid or oil production facilities in the Middle East. Malignant threats are associated with the collapse of a complex system—the electric grid, a human body, the Earth's climate—when it is unintentionally disturbed in some way. Sensible, economically beneficial energy policies have the potential to address both types of problems, enhancing national security and reducing America's contribution to climate change.*

❖

The threat of terrorism rooted in religious or political extremism holds out the possibility of massive damage and loss of life. The scope of death and destruction sought by the perpetrators of this sort of terrorism is something most people find difficult to envision. Terrorism may be considered a "malevolent" rather than a "malignant" problem such as climate change but they share a somewhat surprising connection: The aspects of our energy systems that help create the risk of climate change also create vulnerabilities that terrorists bent on massive destruction are likely to target. We need to be alert to the possibility that although our current circumstances are doubly dangerous, this confluence could give us an opportunity to design a set of changes in our energy systems that will help us deal with both problems.

POSITIVE FEEDBACK LOOPS AND TIPPING POINTS

The climate models agreed upon by the Intergovernmental Panel on Climate Change (IPCC) deal with some, but by no means all, of the warming effects of emissions that can occur as a result of positive feedback loops. This is because climatologists, as scientists, are given to producing testable hypotheses and there are often not enough data to satisfy that requirement for a number of the feedback loop issues. But a number of climatologists have nevertheless assessed the data and offered judgments about the importance of possible feedback effects, even in this century. Positive feedback loops can relatively quickly accelerate climate change to the tipping point, at which it becomes impossible to reverse destructive trends, even with future reductions of greenhouse gas emissions from human activities. Several such positive feedback loops are conceivable in this century, including the risk that freshwater from melting Greenland glaciers would slow the North Atlantic meridional overturning circulation, changing ocean currents and attenuating the Gulf Stream's ability to warm Europe.

Tipping points at which there might be irreversible thawing of Arctic permafrost or the melting and breakup of the West Antarctic and the Greenland ice sheets have such stunning implications they deserve particular attention. Somewhere around a million square miles of northern tundra are underlain by frozen permafrost containing about 950 billion tons of carbon—more than currently resides in the atmosphere.[1] If the permafrost were to thaw, much of this carbon would quickly convert to methane gas. At about one million tons annually, the increase in atmospheric methane content is much smaller than the increase in carbon dioxide (CO_2) content, which weighs in at about 15 billion tons per year.[2] However, a ton of methane affects climate 25 times more powerfully than a ton of CO_2 over a one-hundred-year time horizon.[3] As a result, it would take only 600 million tons of methane to equal the global warming effect of 15 billion tons of CO_2. If this seems like an implausibly large increase in methane emissions, consider that it equates to only one-half of one-tenth of 1 percent of the organic carbon currently preserved in the permafrost (not to mention much larger amounts of frozen methane stored in shallow marine sediments). Therefore, if the permafrost begins to thaw quickly due to the initial linear warming trend we are experiencing today, the climate impact of methane emissions could come to rival that of CO_2 in future decades. Consequent accelerated warming and faster thaw leading to more methane emissions could produce a tipping point beyond which humans no longer control the addition of excess greenhouse gases to the atmosphere, and no options remain under our control for cooling the climate. We don't know the exact point at which this vicious circle would begin, but there are some indications that a substantial permafrost thaw is already under way.[4]

Because of methane's potency, its release could provide a substantial short-term kick to climate change. Such release over a few decades could raise worldwide temperatures by 5°C–6°C (9°F–10.8°F) or more.[5] Another potential feedback loop lurks in the prospect of melting—and sliding—ice sheets in Greenland and West Antarctica. Around one hundred twenty-five thousand years ago, at the warmest point between the last two ice ages, global sea level was 4–6 meters (about 13–20 feet) higher than it is today and global temperature was only about 1°C (1.8°F) higher.[6] Being warmer than Antarctica, Greenland probably provided the initial slug of meltwater to the ocean. However, much of the ice of western Antarctica rests on bedrock far below sea level, making it less stable as sea level rises.[7] When the ice sheet is lubricated by melting where it is grounded, it begins to float and can cause coastal ice shelves to shatter, increasing the rate of ice stream flow into the ocean (ice stream is a region of an ice sheet that moves significantly faster than the surrounding ice).[8] As a result of this action, the West Antarctic Ice Sheet contributed perhaps two meters (six or seven feet) of the additional sea level one hundred twenty-five thousand years ago.

With just 1°C (1.8°F) of warming, therefore, we may be locked into about four to six meters (13–20 feet) of sea level rise.[9] NASA climatologist James Hansen points out that it is not irrational to worry about reaching this tipping point in this century. A catastrophic scenario for future climate change assumes 5°C–6°C (9°F–10.8°F) of warming, which is significantly warmer than conditions 3 million years ago, before the ice ages. At that time, the Earth was 2°C–3°C (3.6°F–5.4°F) warmer and sea level was about 25 meters (82 feet) higher than today.[10] Although the time required for that much sea level rise to occur is probably more than one thousand years, a scenario that assumes two meters (6.6 feet) of sea level rise by the end of this century appears quite plausible.[11]

Robert Zubrin, the author of *Energy Victory: Winning the War on Terror by Breaking Free of Oil*, who is something of a climate change skeptic, highlights one example to illustrate the power of economic growth to affect climate change—a process that could create a climatic tipping point sooner rather than later. The world today has achieved an average GDP per capita comparable to U.S. GDP per capita at the beginning of the twentieth century (about $5,000 in today's dollars).[12] In the twentieth century, world population quadrupled and world economic growth averaged 3.6 percent annually.[13] Even if we assume slower population growth, say a doubling of world population in the twenty-first century, and also a lower growth rate of 2.4 percent—the latter producing a fivefold increase in GDP per capita—unless fuel use per unit of GDP changes substantially, we would see a tenfold increase in carbon dioxide emissions by century's end. This prospect leads even a climate change skeptic such as Zubrin to imagine an extraordinary scenario in which presumably all known and some unknown feedback loops become activated and thus it "only tak[es] a few decades to reach Eocene carbon dioxide atmospheric concentrations of 2,000 ppm"—and certain catastrophe.[14]

To take only one example of the impact of vigorous economic development on carbon dioxide emissions, China is building approximately one large coal-fired power plant per week for the foreseeable future. Rapidly growing developing countries are expected to account for an overwhelming 85 percent of energy-demand growth between 2008 and 2020. China alone represents a third of total growth.[15]

SEA LEVEL RISE AND CHALLENGES TO EXISTING INFRASTRUCTURE

The 2007 Working Group I Contribution to the IPCC's *Fourth Assessment Report* points out that the prospect of climate change and sea level rise coming to a tipping point is particularly troubling because once such a point has been passed, sea level rise will probably continue for centuries.[16] For this reason, James Hansen considers sea level rise as "the big global issue" that will transcend all others in the coming century.[17] Even if the East Antarctic Ice Sheet is not destabilized, the steady melting of the Greenland Ice Sheet together with the perhaps sudden melting of the West Antarctic Ice Sheet holds the potential for some 12 meters (40 feet) of sea level rise.[18] The melting of the East Antarctic shelf would add approximately 25 meters (80 feet); this would mark, in the Antarctic research scholar Peter Barrett's words, "the end of civilization as we know it."[19] Even without a melting of the East Antarctic shelf, civilization would be experiencing an inexorable encroachment of seawater over decades and centuries.

Moreover, humanity would have to face the coastal inundation and related destruction while dealing with substantial disruption of agriculture and food supplies, and resulting economic deprivation, due to changing availability of water—some places more arid, some wetter—and a much smaller percentage of available water would be fresh. Rising sea levels threaten to inundate major portions of cities and wide regions of the U.S. coast from South Texas to West Florida and from East Florida to New York; extensive areas bordering the Chesapeake Bay and most of South Florida and eastern North Carolina; the lower Hudson River Valley; huge shares of the coasts of San Francisco Bay; much of Sydney and all of Darwin, Australia; a large share of Japanese ports; Venice and a major share of coastal Tuscany; the majority of the Netherlands; much of Dublin; a major share of Copenhagen; and the Thames River Valley and the eastern and southern coasts of England.[20] Storm surges would affect people much farther inland and on more elevated coastlines.

Even without considering storm surge, sea level rise in the range of 2 meters (6.6 feet) in this century could have a potentially catastrophic effect on a number of developing countries. According to a February 2007 World Bank policy research working paper, these include particularly Egypt, Vietnam, and the Bahamas and a number of other island nations. It could also have "very large" effects on a number of other states, including China and India. Considering all factors—land area, urban area, population, and so forth—the most affected countries, in addition to those just cited, would be Guyana, Surinam, and Mauritania. Substantial impacts would also occur in Gambia, Liberia, Senegal, Guinea, Thailand, Burma, Indonesia, Taiwan, Bangladesh, and Sri Lanka.

This potential 2-meter (6.6-foot) rise in sea level—together with changed climate, agricultural disruptions and famines, the spread of disease, water scarcity, and severe storm damage—will not occur in a world that is otherwise sustainable and resilient. Many areas are already destabilized. In the Philippines, for example, sea level rise would add to a problem already created by excessive groundwater extraction, which is lowering the land annually by fractions of an inch in some spots to more than a tenth of a meter (three or four inches) annually.[21] The Mississippi Delta has a similar problem with land subsidence. Some of the land south of New Orleans will likely lose about one meter (three feet) of elevation by the end of this century as a result of subsidence.[22] Thus, about six feet (roughly two meters) of sea level rise by the end of the century may well be additive to the substantial lowering of land levels in some areas by such groundwater extraction. And of course the concentration of population in low-lying areas exacerbates the effect of these changes.

Meltwater runoff from mountain glaciers also supplies agricultural and drinking water as well as electricity from hydropower. More than 100 million people in South America and 1–2 billion in Asia rely on glacial runoff for all or part of their freshwater supply. As these glaciers shrink and produce less meltwater they will contribute substantially to the need to emigrate in search of water and arable land. The relevant glaciers are retreating rapidly and some are already virtually gone. This problem is likely to peak within mere decades.[23]

POTENTIAL NATIONAL SECURITY CONSEQUENCES OF CLIMATE CHANGE

In a world that sees a 2-meter (6.6-foot) sea level rise with continued flooding ahead, it will take extraordinary effort for the United States, or indeed any country, to look beyond its own salvation. All of the ways in which human beings have responded to natural disasters in the past could come together in one conflagration: rage at government's inability to deal with the abrupt and unpredictable crises; religious fervor and perhaps even a dramatic rise in millennial end-of-days cults; hostility and violence toward migrants and minority groups at a time of demographic change and increased global migration; and intra- and interstate conflict over resources, particularly food and freshwater.

Altruism and generosity would likely be blunted. In a world with millions of people migrating out of coastal areas and ports across the globe, it will be extremely difficult, perhaps impossible, for the United States to replicate the kind of professional and generous assistance provided to Indonesia following the 2004 tsunami. Even overseas deployments in response to clear military needs may prove very difficult. Nuclear-powered aircraft carriers and submarines might be able to deploy, but aviation fuel or fuel for destroyers and other nonnuclear ships could be unobtainable. Overseas air bases would doubtless be tangled in climatic chaos, and aircraft fuel availability overseas highly uncertain. Further, the navy is likely to be principally involved in finding ways to base, operate, overhaul, and construct ships, as many ports and harbors south of New

York on the East Coast and overseas disappear or become usable only with massive expenditures for protection from the rise in sea levels. Civilians will likely flee coastal regions around the world, including in the United States. The U.S. military's worldwide reach could be reduced substantially by logistics and the demand of missions near our shores.

If Americans have difficulty reaching a reasonable compromise on immigration legislation today, consider what such a debate would be like if we were struggling to resettle millions of our own citizens—driven by high water from the Gulf of Mexico, South Florida, and much of the East Coast reaching nearly to New England—even as we witnessed the northward migration of large populations from Latin America and the Caribbean. Such migration will likely be one of the Western Hemisphere's early social consequences of climate change and sea level rise of this magnitude. Issues deriving from inundation of a large portion of our own territory, together with migration toward our borders by millions of our hungry and thirsty southern neighbors, are likely to dominate U.S. security and humanitarian concerns. Globally as well, populations will migrate from increasingly hot and dry climates to more temperate ones.

On the other hand, extrapolating from current demographic trends, we estimate that there will be fewer than 100 million Russians by 2050, nearly a third of whom will be Muslims. Even a Europe made colder by the degrading of the Gulf Stream may experience substantially increased levels of immigration from south of the Mediterranean, both from sub-Saharan Africa and from the Arab world. Many of Europe's Muslim minorities, including Russia's, are not well assimilated today, and the stress of major climate change and sea level rise may well foster social disruption and radicalization. Russia and Europe may be destabilized, shifting the global balance of power. Northern Eurasian stability could also be substantially affected by China's need to resettle many tens, even hundreds, of millions from its flooding southern coasts. China has never recognized many of the Czarist appropriations of north-central Asia, and Siberia may be more agriculturally productive after a 5°C–6°C (9°F–10.8°F) rise in temperatures, adding another attractive feature to a region rich in oil, gas, and minerals. A small Russian population might have substantial difficulty preventing China from asserting control over much of Siberia and the Russian Far East. The probability of conflict between two destabilized nuclear powers would seem high.

Interactions between climate change and existing infrastructure could create major failures in the systems that support modern civilization. All other systems—from operating telecommunications to distributing food, pumping water, and more—depend on energy. Yet energy systems themselves are vulnerable. Hydroelectric electricity generation may be substantially affected by reduced glacial runoff or by upstream nations diverting rivers in some parts of the world. Nuclear power plant cooling may be limited by reduced water availability. Increased numbers and intensity of storms could interfere with long-distance electricity transmission, already heavily stressed in the United States and elsewhere.

Sea level rise and chaotic weather patterns may interfere with oil production, particularly from sea-based platforms and in parts of the Middle East, and with the operation of large oil tankers. Many U.S. oil refineries are in the Gulf Coast region and thus more vulnerable to disruption by storms than if they were located elsewhere. Hurricane Katrina came very close to shutting down the Colonial Pipeline, the major link from the Gulf Coast to the Eastern Seaboard. In short, the pressures on U.S. society and the world would be significant, and the international community's ability to relieve those pressures seriously compromised. The abrupt, unpredictable, and relentless nature of the challenges will likely produce a pervasive sense of hopelessness.

A MALEVOLENT THREAT: MASS TERRORISM

Our society, our way of life, and our liberty face serious current challenges beyond the infrastructure fragility exacerbated by climate change. The most salient is attack by terrorist groups or an enemy state, or a combination thereof, aimed at massive damage and massive casualties. These are not unintentional "malignant" results of our habitual behavior but are rather "malevolent" and planned carefully by those who want to do far more than many terrorist groups in the past: namely, to destroy our entire civilization and way of life.

An oil-dependent economy presents a panoply of opportunities for highly destructive terrorism. Our transportation is fueled more than 96 percent by petroleum products. Consequently oil has a transportation monopoly in much the same way that, until around the end of the nineteenth century, salt had a monopoly on the preservation of meat —Anne

Korin's excellent analogy. Oil's monopoly creates a litany of vulnerabilities for our society.

Since around two-thirds of the world's proven reserves of conventionally produced oil are in the Persian Gulf region, together with much of the oil industry's international infrastructure, the world's supplies are vulnerable to terrorist attacks such as two already attempted by al Qaeda in Saudi Arabia and emphasized in al Qaeda's doctrine. Some oil states' governments (Iran) are quite hostile today; others (Saudi Arabia) could become so with a change of ruler. A nuclear arms race appears to be beginning between Iran and six Sunni states that have announced nuclear programs "for electricity generation." The United States borrows more than a billion dollars a day at today's prices to import oil, substantially weakening the dollar.

The other major energy sector of our economy, electricity generation and distribution, is also highly vulnerable to attack by terrorists and rogue states. In 2002 the National Research Council published its report on the use of science and technology to combat terrorism. It stated: "The most insidious and economically harmful attack would be one that exploits the vulnerabilities of an integrated electric power grid. ... Simultaneous attacks on a few critical components of the grid could result in a widespread and extended blackout. Conceivably, they could also cause the grid to collapse, with cascading failures in equipment far from the attacks, leading to an even larger long-term blackout."[24] Little has been done to implement the council's 17 detailed recommendations to deal with this, particularly with regard to improving the security of, or even stockpiling spares for, the large transformers at grid substations or effectively protecting the grid's control systems from hacking.

Additionally, the electricity grid has a major vulnerability to an electromagnetic pulse (EMP). In 1962 both Soviet and American atmospheric nuclear tests revealed a troubling phenomenon: Three types of electromagnetic pulses generated at high altitude by nuclear detonations could seriously damage or destroy electronic and electrical systems at as much as 1,610 kilometers (1,000 miles) from the blast. The 2004 report of the U.S. Electromagnetic Pulse Commission pointed out that the detonation of a single nuclear warhead between 40 and 400 kilometers (25 and 250 miles) above the Earth could cause "unprecedented cascading failures of our major infrastructures," primarily "through our electric power infrastructure" crippling "telecommunications...the financial system...means of getting food, water, and medical care to the citizenry...trade...and production of goods and services." The commission noted that states such as North Korea and Iran, possibly working through terrorist groups, might not be deterred from attack (say using a relatively small ship carrying a simple SCUD missile) in the same way as were our adversaries in the Cold War.[25]

❖

Both the malignant and malevolent threats described here are extraordinarily grave. The steps needed to address both, however, have a great deal in common, at least in the important field of energy. As a thought experiment, if we were to summon the counsel of a stereotypical hawk committed to combating terrorism and advancing energy security, and a tree hugger committed to stopping climate change, we would find large areas of common interest. When tasked with this challenge of considering how we generate electricity, fuel transportation, power industry, and operate buildings the hawk and tree hugger could certainly agree that their interests would be served by policies to:

- Improve the energy efficiency of buildings.
- Radically increase the use of combined heat and power.
- Create strong, long-term incentives for small-scale-distributed generation of electricity and heating and cooling.
- Decouple sales from earnings for electric utilities to encourage conservation and grid modernization.
- Give steady and long-term encouragement to the deployment of renewable electricity generation for the grid from wind, solar, hydro, and geothermal sources.

These and other economically rational energy policies would enhance our national security while reducing the United States' contribution to global climate change. Finally, and of central importance, is to move away as quickly as possible from dependence on petroleum-based fuels for transportation. For both climate-change and security reasons this is a matter of extreme urgency. Overall, oil use generates more carbon dioxide than does coal use. A simple change in the kind of plastic used in the fuel lines of vehicles—costing under $100 per car in the manufacturing process—

can make it possible for vehicles to use not only gasoline but also a wide range of alcohol fuels, such as methanol made from (increasingly afford-able) natural gas and from biomass. The use of algal fuels for aviation and of natural gas itself as a fuel for large vehicles, as well as the use of electric-ity generated at least in part from renewables, all hold substantial promise as well. Whether driving on electricity or on methanol, cars would emit much less carbon than when using oil products. We could also largely end our current ridiculous practice of borrowing heavily in order to pay for both sides in our struggle against Islamist terrorism. We will not prevail on climate change or on the key security issues of our time until we drive largely on fuels that are not petroleum-based and make oil products as minor a player in transportation as salt has become in food preservation.

PROGRESS VS. APOCALYPSE

The Stories We Tell Ourselves

JOHN MICHAEL GREER

THE MYTH OF *progress and the myth of apocalypse have deep roots in the collective imagination of the modern world, and few people seem to be able to think about the future without following one story or the other. Both are inappropriate for the futures we are actually likely to encounter. The peak and decline of the world's conventional oil reserves, and the slower exhaustion of coal and other nonrenewable fuels, are setting the industrial world on a trajectory not toward sudden collapse but toward something like the technology and society before the Industrial Revolution.*

❖

Discussions about peak oil and the predicament of industrial society constantly revolve around two completely different, and in fact opposite, sets of assumptions and beliefs about the future. Most people insist that no matter what problems crop up before us, modern science, technology, and raw human ingenuity will inevitably win out and make the world of the future better than the world of today. A sizeable minority of people, however, insist that no matter what we do, some overwhelming catastrophe will soon bring civilization suddenly crashing down into mass death and a Road Warrior future.

These competing narratives reflect the hidden presence of myth. Many people nowadays think only primitive people believe in myths, but myths dominate the thinking of every society, including our own. Myths are the stories we tell ourselves to make sense of our world. Human beings think with stories as inevitably as they see with eyes and walk with feet, and the most important of those stories—the ones that define the nature of the world for those who tell them—are myths.

According to the *myth of progress*, all of human history is a grand tale of human improvement. From the primitive ignorance and savagery of our cave-dwelling ancestors, people climbed step by step up the ladder of progress, following in the wake of the evolutionary drive that raised us up from primeval slime and brought us to the threshold of human intelligence. Ever since our ancestors first became fully human, knowledge gathered over the generations made it possible for each culture to go further, become wiser, and accomplish more than the ones that came before it. With the coming of the scientific revolution three hundred years ago, the slow triumph of reason over nature shifted into overdrive and has been accelerating ever since. Eventually, once the last vestiges of primitive superstition and ignorance are cast aside, our species will leap upward from the surface of its home planet and embrace its destiny among the stars.

Conversely, the *myth of apocalypse* suggests that human history is a tragic blind alley. People once lived in harmony with their world, each other, and themselves, but that golden age ended with a disastrous wrong turning and things have gone downhill ever since. The rise of vast, unnatural cities governed by bloated governmental bureaucracies, inhabited by people who have abandoned spiritual values for a wholly material existence, marks the point of no return. Sometime soon the whole rickety structure will come crashing down, overwhelmed by sudden catastrophe, and billions of people will die as civilization comes apart and rampaging hordes scour the landscape. Only those who abandon a corrupt and doomed society and return to the old, true ways of living will survive to build a better world.

Both these myths have deep roots in the collective imagination of the modern world, and very few people seem to be able to think about the future at all without following one story or the other. It would be hard to find any two narratives less appropriate, though, for the futures we are actually likely to encounter. Both of them rely on assumptions about the world that won't stand up to critical examination.

The faith in progress, for example, rests on the unstated assumption that limits don't apply to us, since the forward momentum of human progress automatically trumps everything else. If we want limitless supplies of energy badly enough, the logic seems to be, the world will give it to us. Of course the world *did* give it to us—in the form of unimaginably huge deposits of fossil fuels stored up over hundreds of millions of years through photosynthesis—and we wasted it in a few centuries of fantastic extravagance. The lifestyles we believe are normal are entirely the product of that extravagance.

Nor is the past quite so much of a linear story of progress as the folklore of the industrial age would have it. The lives of peasants, priests, soldiers, and aristocrats in Sumer in 3000 BCE differed only in relatively minor details from those of their equivalents in Chou dynasty China fifteen centuries later, Roman North Africa fifteen centuries after that, or medieval Spain another fifteen centuries closer to us. The ebb and flow of technologies before the modern period had little impact on daily life, because without cheap abundant energy to power them, it was more efficient and economical to rely on human labor with hand tools.

This stable pattern changed only when the first steam engines allowed people to begin tapping the fantastic amounts of energy hidden away within the Earth. That and nothing else brought the industrial world into being. For thousands of years before that time, everything else necessary for an industrial society had been part of the cultural heritage of most civilizations. Renewable energy sources? Wind power, waterpower, biomass, and muscle power were all used extensively in the pre-industrial past without launching an industrial society. Scientific knowledge? The laws of mechanics were worked out in ancient times, and a Greek scientist even invented the steam turbine two centuries before the birth of Christ; without fossil fuels it was a useless curiosity. Human resourcefulness and ingenuity? It's as arrogant as it is silly to insist that people in past ages weren't as resourceful and ingenious as we are.

Fossil fuel energy, and only fossil fuel energy, made it possible to break with the old agrarian pattern and construct the industrial world. Unless some new energy source as abundant as fossil fuels comes online before today's extravagant lifestyles finish burning through the resources that made the industrial age possible, we will find ourselves back in the same world our ancestors knew, with the additional burdens of a huge surplus population and an impoverished planetary biosphere to contend with. Combine these constraints with the plain hard reality that the fossil fuels that made industrialism possible won't be there anymore, and the myth of perpetual progress becomes a mirage.

Believers in apocalypse, for their part, insist that the end of industrial civilization will be sudden, catastrophic, and total. That claim is just as hard to square with the realities of our predicament as the argument for perpetual progress. Every previous civilization that has fallen has taken centuries to collapse, and there's no reason to think the present case will be any different. The resource base of industrial society is shrinking but it's far from exhausted, the impact of global warming and ecological degradation build slowly over time, and governments and ordinary citizens alike have every reason to hold things together as long as possible.

The history of the last century—think of the Great Depression, the Second World War, and the brutal excesses of Communism and Nazism—shows that industrial societies can endure tremendous disruption without dissolving into a Hobbesian war of all against all, and people in hard times are far more likely to follow orders and hope for the best than to turn into the rampaging, mindless mobs that play so large a role in survivalist fantasies these days.

The Hollywood notion of an overnight collapse makes for great screenplays but has nothing to do with the realities of how civilizations fall. In the aftermath of Hubbert's peak, fossil fuel production will decline gradually, not simply come to a screeching halt, and so the likely course of things is gradual descent rather than free fall, following the same trajectory marked out by so many civilizations in the past. This process is not a steady decline, either; between sudden crises come intervals of relative stability, even moderate improvement; different regions decline at different paces; existing social, economic, and political structures are replaced, not with complete chaos, but with transitional structures that may develop considerable institutional strength themselves.

Does this model apply to the current situation? Almost certainly. As oil and natural gas run short, economies will come unglued and political systems disintegrate under the strain. But there's still oil to be had—the Hubbert curve is a bell-shaped curve, after all, and assuming the peak came around 2010, the world in 2040 will be producing about as much oil as it was producing in 1980. With other fossil fuels well along their own Hubbert curves, nearly twice as many people to provide for, and a global economy dependent on cheap abundant energy in serious trouble, the gap between production and demand will be a potent driver of poverty, spiralling shortages, rising death rates, plummeting birth rates, and epidemic violence and warfare. Granted, this is not a pretty picture, but it's not an instant reversion to the Stone Age either.

THE RELIGION OF PROGRESS

The myth of progress, like the belief that everyone creates their own reality, raises expectations that the real world in an age of diminishing resources simply isn't able to meet. As the gap between expectation and experience grows, so, too, does the potential for paranoia and hatred. Those who cling to faith in progress are all too likely to go looking for scapegoats when the future fails to deliver the better world they expect. The sheer emotional power of the myth of progress makes this a difficult trap to avoid. The claim that progress is inevitable and good has become so deeply woven into our collective thinking that many people simply can't get their minds around the implications of fossil fuel depletion, or for that matter any of the other factors driving the contemporary crisis of industrial civilization. All these factors promise a future in which energy, raw materials, and their products—including nearly all of our present high technology—will all be subject to ever-tightening limits that will make them less and less available over time. Thus we face a future of regress, not progress.

The problem here is that regress is quite literally an unthinkable concept these days. Suggest to most people that progress will soon shift into reverse, and that their great-great-grandchildren will make do with technologies not that different from the ones their great-great-grandparents used as the industrial age gives way to the agrarian societies of a deindustrial future, and you might as well be trying to tell a medieval peasant that heaven with all its saints and angels isn't there anymore. In words made famous a few years ago by Christopher Lasch, progress is our "true and only heaven"; it's where most modern people put their dreams of a better world, and to be deprived of it cuts to the core of many people's view of reality.

Even those who reject the myth of progress in favor of the myth of apocalypse draw most of their ideas from the faith they think they've abandoned. Like Satanists who accept all the presuppositions of Christian theology but root for the other side, most of today's believers in apocalypse swallow whole the mythic claim that progress is as inevitable as a steamroller; it's simply that they believe the steamroller is about to roll its way off a cliff. The suggestion that the steamroller is in the process of shifting into reverse, and will presently start rolling patiently back the way it came, is just as foreign to them as it is to believers in progress.

It's not going too far, I think, to call belief in progress the established religion of the modern industrial world. In the same way that Christians have traditionally looked to heaven and Buddhists to nirvana, most people look to progress for their hope of salvation and their explanation for why the world is the way it is. To believers in the religion of progress, all other human societies are steps on a ladder that lead to modern industrial civilization. Progress means our kind of progress, since our civilization has by definition progressed further than anyone else's, and the road to the future thus inevitably leads through us to something like our society, but even more so. All of these claims are taken for granted as self-evident truths by most people in today's industrial societies. Not one of them has a basis in logic or evidence; like the doctrine of the Trinity or the Four Noble Truths, they are statements of faith.

The religion of progress has a central role in driving the predicament of industrial civilization, because the dead end of dependence on rapidly depleting fossil fuels can't be escaped by going further ahead on the path we've been following for the last three centuries or so. Almost without exception, the technological progress of the industrial age will have to shift into reverse as its foundation—cheap abundant energy—goes away, and most of the social and cultural phenomena that grew out of fossil fuel–powered technology will go away as well. The peak and decline of the world's oil reserves is the first step in this process, and the slower exhaustion of coal and other nonrenewable fuels will complete it, setting the industrial world on a trajectory that will most likely lead to something like the technology and society it had before the industrial revolution began in the first place.

When the parade of wonders grinds to a halt, the impact on deindustrializing cultures may be immense. If progress is indeed the unrecognized religion of the industrial world, the failure of its priests to produce miracles on cue could plunge many people into a crisis of faith with no easy way out. The recent vogue for conspiracy theories and apocalyptic visions of the future strike me as two clear signs that people are beginning to turn their back on the religion of progress and seek their salvation from other gods. Neither pursuing scapegoats nor waiting for redemption through catastrophe are particularly productive as ways of dealing with the transition to the deindustrial age, but both of them offer a great deal of emotional consolation, and it's likely to take more than the usual amount of clearheadedness to avoid them in the difficult times we are likely to encounter in the near future.

Peoples of the past, stripped of their traditional faith in one way or another, have responded in many ways. Some have launched revitalization movements to renew the old faith or to revive some older vision of destiny, some have embraced newly minted belief systems or traditions imported from distant lands, and some have simply huddled down into themselves and died. We have begun to see examples of each of these reactions in the modern industrial world. Which of them turns out to be most common may have drastic effects on the way the twilight of the industrial age works out, because the stories we tell ourselves will have an immense impact on the world we create at the end of the industrial age.

Part Three

THE LANDSCAPE OF ENERGY

INTRODUCTION

A Tour of the Energy Terrain

DAVID MURPHY

The next decades will witness a global battle between geologic depletion and technological advancement, as modern society demands
ever-increasing quantities of energy from an aging fossil fuel supply and a
nascent renewable energy sector.

The fossil fuel industry is already straining to deliver increasing quantities of energy from geologic reservoirs that are old and depleting quickly,
or from new ones that are more energy intensive and expensive to develop.
Meanwhile, companies and governments alike are rushing to develop
renewable energy technologies that can compete on a cost basis with fossil
fuels. Renewable energy optimists believe that once these price barriers are
defeated (through technological advancement or market manipulation), both
our energy concerns and the environmental problems associated with the
present-day energy industry will be vastly reduced.

The outcome of this battle can only be analyzed retrospectively, but
as we search for oil under the Arctic ice cap and coat the deserts with solar
panels, we can anticipate that it will extend across all landscapes thought
to be energetically bountiful.

One thing is clear: When it comes to energy, there is no free lunch.
It would be foolish to assume that transitioning to renewable energy will
solve all of our energy and environmental problems. Transitioning to
renewables will certainly diminish ecological impacts in many ways, but it
will also have new—and mostly unknown—consequences. For example,
both solar and battery technology in their current iterations depend on
rare metals and other natural resources that are unevenly distributed
around the world. A full-scale switch to renewable energy may merely
supplant one dependency for another.

It would be wise to approach our energy future with two related
thoughts in mind: first, the precautionary principle, and second, the law of
unintended consequences. Using that perspective, this section of the book

reviews the major energy resources and their transportation methods. We consider the current status of each resource as well as any other major concerns, environmental and otherwise, that may exist.

Of course the energy economy is constantly in flux; the following overview of the energy landscape can necessarily provide only a snapshot in time. It is intended to offer the reader a foundation of understanding about the current energy mix, building on the "energy literacy" series in part one.

THE LANDSCAPE OF ENERGY

Our global system of energy production and consumption faces systemic challenges that have been centuries in the making—and that no one technology or fantastic new resource will solve. As the figures on the following pages illustrate, we face major obstacles to continuing our recent energy bonanza through the twenty-first century and beyond.

The global economy requires a massive, uninterrupted flow of energy every single day, the vast majority in the form of fossil fuels. These non-renewable energy resources are constantly depleting, so we must continually find and successfully exploit new sources so that tomorrow we can meet the same demand as we did today—plus additional resources for the world's growing population and economies.

But the transition to new sources will be far from seamless: The physical infrastructure of modern society is designed to run primarily on cheap, powerful conventional fossil fuels. Unconventional fossil fuels come with greater economic and environmental costs, significantly reducing their energy benefit, while renewables fall short at matching many of the characteristics we so value in fossil fuels.

An analysis of the available energy resources shows that there are no easy answers.

CHARACTERISTICS OF ENERGY RESOURCES

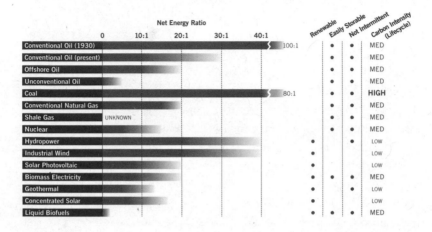

An energy resource is unhelpful if it requires nearly as much energy to produce as it provides to society. The net energy ratio gives us an approximate indication of this relationship. Similarly, an energy resource is worthless if we can't use it the way we need it. The world's infrastructure for transportation and commerce was built for oil and coal power in large part because these resources are relatively easy to store and transport, and can be used at will. Most renewables lack these attributes. The environmental impact of a resource—including but not limited to its carbon intensity—is key to its long-term utility, and the main argument against coal as a baseload power source. (Data: D. Murphy)

BUSINESS–AS–USUAL ASSUMPTIONS

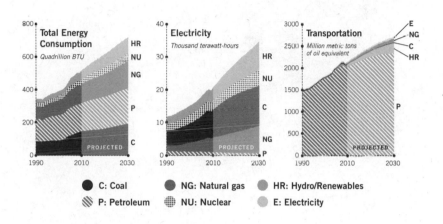

Governments and the oil industry generally assume that market demand and improving technologies will continue to drive increased fossil fuel production. The above forecasts are based on business-as-usual economic growth models and assume that the resource will be exploitable given the right market conditions. Unfortunately, such predictions tend to ignore constraints that cannot necessarily be solved by higher energy prices (e.g., lack of remaining sites suitable for large-scale hydropower, or the long-term nuclear waste problem). Worse yet, even in these rosy scenarios renewable energy is expected to play only a minor role in total energy consumption, in electricity generation, and especially in transportation. (Data: EIA, BP)

WORLD ENERGY USE

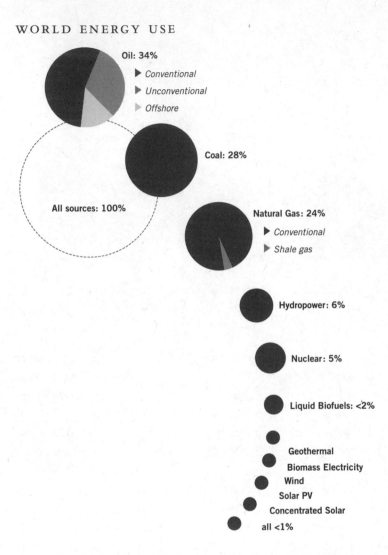

Oil: 34%
▶ Conventional
▶ Unconventional
▶ Offshore

Coal: 28%

All sources: 100%

Natural Gas: 24%
▶ Conventional
▶ Shale gas

Hydropower: 6%

Nuclear: 5%

Liquid Biofuels: <2%

Geothermal
Biomass Electricity
Wind
Solar PV
Concentrated Solar
all <1%

A graphic look at world energy consumption suggests the scale of effort that will be required to dramatically reduce fossil fuel use and ramp up renewables. Considering the constraints already limiting large-scale renewable energy development (most sites suitable for large hydropower have been developed; the large land footprints for some industrial solar, wind, and biomass plants have stoked intense opposition), it's hard to see how the balance of world energy sources will change significantly without serious conservation efforts—that is, efficiency and curtailment. (Data: EIA, C. Campbell)

GOING DEEPER

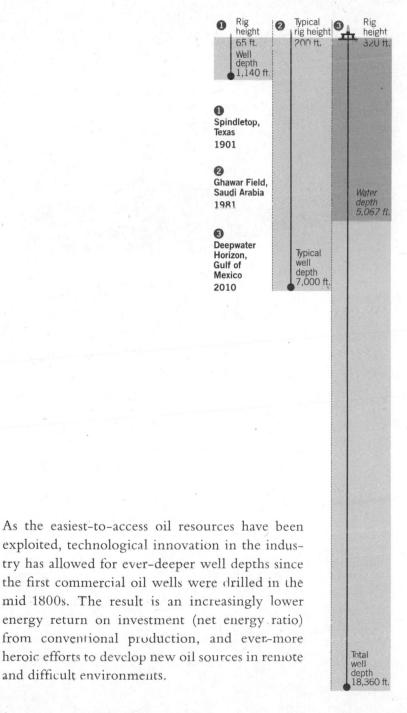

As the easiest-to-access oil resources have been exploited, technological innovation in the industry has allowed for ever-deeper well depths since the first commercial oil wells were drilled in the mid 1800s. The result is an increasingly lower energy return on investment (net energy ratio) from conventional production, and ever-more heroic efforts to develop new oil sources in remote and difficult environments.

THE REALITIES OF DECLINE: CONVENTIONAL OIL

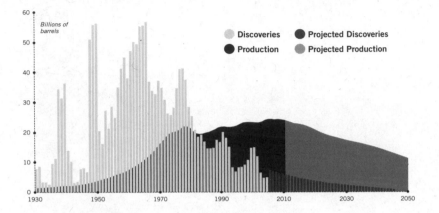

Discoveries of conventional oil peaked in the 1960s and have since slowed to a trickle. Thus it is widely expected that production of conventional oil will soon enter terminal decline—indeed, production has essentially leveled off since 2005. Discoveries of unconventional oil (such as tar sands and shale oil) are still on the rise, but they too will eventually hit their peaks, as will coal and natural gas. (Data: ASPO, ExxonMobil)

CONVENTIONAL OIL

Oil is the lubricant of modern civilization, and a major driver of the global eco-social crisis, manifest in unraveling ecosystems and loss of traditional cultures. By fueling an insatiable industrial-growth economy, oil's aggregate damage to ecological, cultural, and political integrity is incalculable.

Inexpensive and abundant supplies of oil and other fossil fuels have been used to support virtually every aspect of economic life in the over-developed countries. Unparalleled as a transport fuel, more than 30 billion barrels of oil are consumed globally every year. The United States uses roughly 7 billion barrels annually, or 22 percent of world oil consumption. Oil is easily stored and transported and is extremely energy dense. A single liter of oil has an amount of energy equivalent to a human performing hard labor for hundreds of hours.

Oil is the residue of ancient marine plankton transformed by heat and pressure over millions of years. After an underground reservoir is discovered, drilling and pumping bring the oil to the surface where it is sent to refineries. At refineries, which are among the worst of all industrial polluters, the crude is processed into heating oil, kerosene, aviation fuel, diesel, gasoline, etc. Other products derived from oil range from cosmetics to plastics to asphalt. We even "eat" oil: The energy inputs that undergird industrial agriculture, including synthetic pesticides, largely come from oil. By some estimates, our food system uses more than seven calories of energy for every calorie of food consumed.

Finding, extracting, refining, and burning oil produces enormous land, water, and air pollution, including greenhouse gas emissions. Conventional oil field development creates massive networks of roads and pipelines that destroy and fragment wildlife habitat, and a vast global distribution network essentially guarantees ongoing oil spills, from leaking car motors and storage tanks to occasional Exxon Valdez– and Deepwater Horizon–scale disasters.

The oil age began in the 1850s when wells were drilled in Canada and the United States, launching a century and a half of explosive economic and population growth. The first oil to be found and produced was, naturally, the easiest to extract and therefore the cheapest; it also happened to be of superior quality, generally offering a net energy ratio of well over 25:1. Worldwide discoveries of this "conventional" oil peaked in the 1960s, however, and worldwide production has flattened over the last decade, despite record-high prices. It is widely accepted that the age of "easy" oil is coming to a close. Society is turning toward deepwater offshore oil, tar sands, oil shales, and other more challenging resources to meet ever-growing global demand for oil.

Key Limiting Factor *Discoveries peaked in the 1960s; production in terminal decline.*

Net Energy Ratio

| 0 | 10:1 | 20:1 | 30:1 |

OFFSHORE OIL

Offshore drilling, particularly in deepwater, is one of the frontiers for oil exploration in what security expert Michael Klare calls "the era of extreme energy." The deeper the water and the farther from land, the more complex the production challenges, leading to increased risk of catastrophic accidents.

Significant offshore oil development has occurred in many parts of the world including the Gulf of Mexico, along the coasts of Newfoundland and Labrador in Canada, coastal Mexico, the Gulf of Guinea off the coast of West Africa, the North Atlantic, and coastal Brazil. Offshore drilling has been going on for more than a century, with the first saltwater operations organized in 1896 off the coast of Santa Barbara, California. Offshore platforms are responsible for roughly 22 percent of oil production and 12 percent of natural gas production in the United States. As terrestrial reserves are depleted and drilling technology improves, offshore production is expected to increase.

In shallow water, offshore drill rigs are often anchored to the sea floor. In deeper water, floating platforms fixed by chains to the sea floor allow drilling in water depths of 10,000 feet or more. Drilling platforms are expensive to construct and the additional distance of drill pipe that must pass through the water column between the sea bottom and drill platform adds to the difficulty and cost of bringing oil to the surface. Logistical expenditures associated with getting supplies and highly trained crews to and from platforms also add greatly to the cost, in both money and energy, and are one of the reasons that deepwater drilling inherently has a lower net energy ratio (energy return on energy investment) than conventional oil production on land.

Offshore drilling magnifies the risk from accidents because ocean currents can easily distribute any spilled oil great distances. The 2010 Deepwater Horizon explosion in the Gulf of Mexico caused an estimated 4.9 million barrels of oil to leak before the well was capped. As offshore oil production expands into Arctic waters, the ecological risks associated with an oil spill are magnified because cold temperatures hinder the biological breakdown of oil. Moreover, Arctic rescue, repair, and cleanup operations would be severely complicated by the remote distances and harsh weather conditions. Beyond the environmental and human risks, offshore development has aesthetic impacts. Every additional drilling platform industrializes the ocean, compromises beauty, and degrades the wilderness character of the marine environment.

Key Limiting Factor *Production challenged by extreme environments and significant technological complexity.*

Net Energy Ratio

0   10:1   20:1

UNCONVENTIONAL OIL

Unconventional liquid fuels are more polluting than conventional oil. Wholesale development of tar sands, shale oil, kerogen, and other unconventional oil resources will likely doom humanity's attempt to rein in greenhouse gas emissions and almost certainly tip the world toward climate chaos.

The liquid fuels that can be produced from tar sands, oil shale (or kerogen), shale oil formations, and coal are generally lumped under the term "unconventional oil." Coal-to-liquids technology has been known for decades—Nazi Germany used it during World War II—but has never been economically competitive with conventional crude oil and is not projected to grow into a major energy source. Tar sands and shale oil, however, have received significant attention and investment in recent years, with promoters claiming that these sources could soon make North America oil-independent.

Tar sands production is already commercially viable (with oil prices over about $60 per barrel) and has increased rapidly over the past decade. Tar sands contain a viscous substance called "bitumen" (similar to very heavy crude oil) tied up in sand or clay. The greatest known reserves are in Alberta, Canada. Typically, tar sands are strip-mined and the bitumen cooked out. Natural gas has been the primary energy source to do this, making the greenhouse gas footprint of tar sands oil far larger than that of conventional oil. The process also causes deforestation and leaves toxic lakes of wastewater slurry. Some of the environmental impacts associated with tar sands may decrease in the future as the industry adopts in situ extraction methods that are claimed to make the environmental impacts comparable to conventional oil operations.

Oil shales are widely distributed around the world but the largest deposits are in the western United States. Roughly two-thirds of known oil shales, containing an estimated 1.7 trillion barrels of oil equivalent, are found in Wyoming, Utah, and Colorado. In oil shales, the hydrocarbons are in the form of kerogen, a precursor to oil that has not been heated long enough by geological processes to become oil or natural gas. Oil shales can be converted to oil and natural gas through a variety of techniques, most of which require heating the rock above 600°F (315°C).

Shale oil, also known as "tight oil," is high quality light crude that is trapped in rock formations of low permeability. The horizontal drilling and hydraulic fracturing ("fracking") techniques used for shale gas production have recently been successfully applied to shale oil production, primarily in North Dakota and Texas. Costs are high and from early data it appears that, like shale gas wells, shale oil wells may have fairly low lifetime productivity.

Some estimates put the total resource in tar sands, oil shales, and shale oil at 6 trillion barrels, more than known conventional oil reserves. But extracting oil from these sources has proven to be much more energy-intensive and damaging than conventional oil production, and poses a grave threat to both local ecosystems and the global climate. The enormous technical complications of unconventional oil suggest that it will be extremely challenging to ramp up their production to fully replace declining conventional oil resources at the scale and rate needed.

Key Limiting Factor *Significant energy, water, and infrastructure investments required.*

Net Energy Ratio �

0   5:1

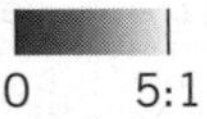

NATURAL GAS

The "clean" fossil fuel, natural gas is often mistakenly thought to have very little environmental impact. But newer extraction methods and the sheer quantity of natural gas consumed make it one of the largest greenhouse gas contributors globally.

Oil's sibling is natural gas. Methane, the primary constituent of natural gas, is formed by the breakdown of organic material. In landfills, the rapid breakdown of organic material in the absence of oxygen creates a mix of gases that includes methane; under the Earth's crust the breakdown of prehistoric plankton forms both oil and natural gas. To form natural gas, the plankton is simply "cooked" at higher temperatures and pressures for longer periods of time than for oil, breaking the molecules into shorter chains of carbon atoms. But since the pressure and temperature can vary even within one hydrocarbon reservoir, oil and natural gas are often found together.

Global natural gas production equals 3,139 billion cubic meters annually, which is the energetic equivalent of 21 billion barrels of oil per year—roughly two thirds the energy content of all the oil produced in the world. Global proved reserves of conventional natural gas are distributed widely, with the largest shares belonging to Russia (24 percent), Iran (16 percent), and Qatar (14 percent). The United States is currently the world's top natural gas producer, followed by Russia.

Natural gas is a highly coveted resource; it can have a high energy density (when pressurized into a liquid form), and it produces fewer greenhouse gas emissions at the burner tip than oil and coal. While natural gas is traded globally, its transport by sea in the form of liquefied natural gas (LNG) requires significant specialized infrastructure, making it more of a regional resource compared to oil. In the United States, natural gas is used prominently in the electricity sector to meet peak demand, as well as in a variety of functions in the industrial and manufacturing sectors. Natural gas also serves as the main heating and cooking fuel for much of the United States and world.

Conventional natural gas has the lowest greenhouse gas emissions per unit of energy of all the fossil fuels, but since it is used in such high quantities it accounts for over 20 percent of U.S. carbon dioxide emissions. Hydraulic fracturing (or "fracking") is now being used to produce harder-to-access shale gas deposits, and this may increase considerably the greenhouse gases per unit of energy from natural gas. Recent studies suggest that this increase may nullify any potential savings in greenhouse gas emissions from burning natural gas instead of oil or coal. Some climate and energy experts argue, however, that with strong regulation of the industry, including standards on preventing unburned methane leakage systemwide, significant greenhouse gas reductions can be achieved from burning natural gas rather than coal.

Key Limiting Factor *Largely a regional fuel because of overseas transport complications.*

Net Energy Ratio

0	10:1	20:1

SHALE GAS

Whether the current boom in hydraulic fracturing ("fracking") for shale gas is a short-lived bubble or a natural gas revolution, it threatens to increase pollution, destroy habitat, and keep America hooked on a "bridge fuel" to nowhere.

Fracking fluid is a mixture of chemicals, water, and sand that is injected under extreme pressure into a shale formation, opening cracks in the shale that release the natural gas trapped in the rock. Shale gas deposits are widespread in North America, Europe, Asia, and Australia. In the United States, the Marcellus Shale running from New York to West Virginia is the epicenter of the shale gas rush. The Barnett Shale of Texas, the Hayesville Shale in Louisiana, and the Fayetteville Shale of Arkansas are other important U.S. shale gas deposits. Many land-owners and rural communities see shale gas development as an economic windfall.

U.S. conventional natural gas production peaked in 2001 and was thought to be in terminal decline before the fracking boom, which has reversed the production trend and prompted the U.S. Energy Information Administration to increase its estimate of recoverable domestic natural gas reserves. In 2009 shale gas provided 14 percent of U.S. natural gas supplies and is officially projected to grow to 46 percent by 2035. But some energy experts think that is unlikely because of high per-well costs and steep decline rates in shale gas wells.

As with conventional gas production, shale gas development entails clearing land for drill pads, access roads, and pipelines. But unlike conventional gas production, fracking consumes copious quantities of water—up to several million gallons per well—which may lead to intense competition for water in more arid parts of the country. Furthermore, the drilling fluid and wastewater that remain after fracking are full of largely undisclosed chemicals, some toxic, that may contaminate groundwater if spilled or leached into nearby streams.

While natural gas has been viewed as the least-polluting fossil fuel, analyses of the life-cycle greenhouse gas emissions of shale gas production by some scientists have suggested that it is not much better than coal power, largely due to increased methane release during drilling and transmission. If confirmed, this undermines the idea that shale gas is a less polluting "transition fuel" toward renewables.

Key Limiting Factor *Young sector with uncertainties about long-term productivity.*

Net Energy Ratio UNKNOWN

COAL

It launched the Industrial Revolution, birthed the modern energy economy, and put civilization on a trajectory toward hypercomplexity and exponential growth. And now that remarkable rock that burns is helping cook the planet.

Coal is the fossilized remains of ancient plants that accumulated on the bottom of shallow water bodies before being buried by sediment during the Carboniferous and Permian periods some 363–245 million years ago. There are four basic kinds of coal, which vary in their energy density due mainly to carbon content: Anthracite has the highest energy content, followed by bituminous, sub-bituminous, and lastly lignite (also called "brown coal").

Large coal deposits are located in the United States, Russia, China, and Australia. Globally, coal use is increasing rapidly, particularly in China, which burns roughly half of the world's annual coal production. Consequently, China has become the world's leading emitter of greenhouse gas pollution. The United States, sometimes called the "Saudi Arabia of coal," has roughly 29 percent of the world's coal reserves, which are used to provide nearly half of U.S. electricity generation. There are more than 600 coal-fired electricity generating facilities in the United States, with dozens of new ones either under construction or seeking permits. Clean energy activists have successfully blocked more than a hundred proposed coal plants in recent years, and low natural gas prices are leading utilities to close some older, heavily polluting coal plants or convert them to natural gas.

Coal is relatively easy to mine, transport, and store and is perceived to be a cheap source of energy. It is very expensive, however, if the associated ecological and public health costs are considered. A 2011 study published in the *Annals of the New York Academy of Sciences* attempted a full life-cycle accounting of coal's public health and environmental costs; it estimated that these "externalities" may exceed $500 billion annually in the United States alone. Coal combustion can also release large quantities of toxins including mercury, lead, arsenic, and sulfur dioxide. Particulates released by coal burning are also a major pollutant and are blamed for tens of thousands of heart attacks and premature deaths in the United States each year.

Globally, coal burning is responsible for more than 40 percent of human-caused carbon dioxide emissions, and thus is a key factor in climate change. Coal mining, processing, and burning also produce vast amounts of liquid and solid pollution including coal combustion ash, which has caused contamination in dozens of states according to EPA and conservation group studies.

Recent efforts have sought to clean up coal's image with promises of carbon capture and storage (CCS) and smokestack scrubbers. But there are serious doubts about the scalability of CCS technology, and both sequestration and scrubbers require additional energy—meaning yet more coal must be burned to generate the equivalent energy delivered to consumers.

Key Limiting Factor *Worst polluter of the fossil fuels.*

Net Energy Ratio

0	10:1	20:1	30:1	40:1	80:1

NUCLEAR

Nuclear plants can generate large quantities of relatively dependable base-load electricity but are tremendously costly to build, produce dangerous radioactive waste, and present an attractive target for terrorism. Other options cost less and produce no deadly long-lived waste, for which there still is no permanent storage option in the United States.

There are more than 400 nuclear power plants currently operating in 31 countries around the world. Roughly 13–14 percent of the world's electricity comes from nuclear power. The United States produces the most nuclear energy of any country, although this accounts for only 19 percent of its electricity. France, by comparison, generates about half as much power from nuclear energy, but that amount represents almost 80 percent of its electricity production, the highest proportion in any nation.

Proponents argue that nuclear power is a safe, carbon-free source of power, and that it presents a green alternative to dirty, climate-killing coal. This claim does not hold up well to critical scrutiny. Although nuclear plants do not emit carbon dioxide while heating water to run a steam turbine (as coal-burning power plants do), life-cycle analysis of nuclear power shows that the entire process emits significant greenhouse gases. Deforestation and mining to procure uranium, nuclear plant construction with massive amounts of steel and concrete, and decommissioning and waste storage responsibilities that stretch thousands of years into the future all are significant greenhouse gas contributors.

High-profile accidents including the Chernobyl, Three Mile Island, and Fukushima Daiichi reactor meltdowns have periodically focused world attention on the potential for catastrophic breakdown of these highly complex systems. While major accidents are rare, "near misses" occur more frequently, and small releases of radioactivity are common. After many decades of trying to solve the waste disposal problem, the United States still has no permanent repository for high-level nuclear waste. Moreover, nuclear plants are an obvious target for terrorists, and a civilian nuclear industry can be used by rogue nations to develop nuclear weapons capability. Finally, a key objection to nuclear power is its tremendous cost: Without government support including loan guarantees and insurance underwriting, private capital markets in the United States would not finance new nuclear plant construction.

Key Limiting Factors *No good solution yet available for extremely long-lived radioactive wastes; not economically viable without government underwriting.*

Net Energy Ratio

0   10:1   20:1

HYDROPOWER

Humans have long harnessed the kinetic energy of falling water—but it was the development of modern construction methods that allowed for the rise of megadams around the globe. Large-scale hydropower is lauded as a greenhouse gas-free energy source, but it effectively kills wild rivers, dramatically altering ecosystem structure and function to generate electricity.

Like windpower, hydropower has a long history of use around the world. Ancient societies used watermills for grinding grain and other mechanical needs. Hydropower is now the largest and lowest-cost source of renewable energy in the world, with some 777 gigawatts of installed capacity. China's Three Gorges Dam is the single largest electricity-generating facility in the world, producing 20 gigawatts of power—more than 20 times the size of the average coal power plant.

The first hydroelectric dams were installed in the United States in the late 1800s; today hydropower accounts for 6 percent of all U.S. electricity generation and 60 percent of the electricity production from all renewable resources. Hydropower is considered one of the least-polluting energy sources because of its low greenhouse gas emissions, but it does have serious ecological impacts. Damming a river can completely alter the natural ecosystem by flooding the upstream portion and altering flow rates and natural silt deposition downstream. The resulting habitat fragmentation, loss of water quality, and changes in species diversity may put increased pressure on vulnerable species. In the Pacific Northwest, for example, the vast network of hydroelectric dams installed on the Columbia River system in the twentieth century decimated regional salmon populations, wiping out entire runs and endangering many others.

The future growth potential of hydropower in most developed countries is limited. More than 45,000 large dams already degrade rivers across the Earth. Since much of the hydropower infrastructure in the United States is old, there are efficiency improvements that can be realized by upgrading dams, but most major rivers that have potential for producing electricity are already dammed. Small-scale hydropower ("micro-hydro") and so-called "run of the river" technologies that generate power without dams or impoundments can be ecologically benign and a useful part of regional distributed energy efforts; their total potential generating capacity, however, is modest. Other emerging hydropower technologies, including tidal and wave power, have not yet proven commercially viable and are far inferior (in terms of cost and power generation) when compared to traditional hydropower.

Key Limiting Factors *Best sites already developed; megadams destroy natural hydrology of river systems, may imperil species and displace human communities.*

Net Energy Ratio

0	10:1	20:1	30:1	40:1

GEOTHERMAL

Geothermal energy utilizes the heat produced by the Earth's core to create electricity and to heat homes. Only certain locations have the appropriate mix of resource availability and high population densities to make this resource substitutable for fossil fuels.

Geothermal energy is naturally vented at the Earth's surface in the form of volcanoes, geysers, and hot springs. Large amounts of geothermal energy are vented at the intersections of tectonic plates as well, such as along the Pacific Rim. Geothermal energy can be used to produce electricity by either harnessing steam directly from geothermal resources or by using hot geothermal water to produce steam to run a turbine. It is also used to heat and cool buildings.

With 15 gigawatt-hours of electricity generated in 2010 from more than 70 power plants, the United States is the world's leading producer of electricity from geothermal sources—but this amounts to less than 1 percent of total nationwide electricity consumption. Few countries produce a significant share of electricity from geothermal sources; only Iceland, El Salvador, and the Philippines use it to generate more than 15 percent of their electricity. Since the footprint of a geothermal plant is fairly small, most of the environmental damage that comes from geothermal energy production is associated with the construction of the facility and its related transmission infrastructure.

Unlike wind and solar energy, which are intermittent sources of power, geothermal energy is consistent, and thus is one of the only renewable energy technologies that substitutes well for coal generation. (Coal power plants take hours to become hot enough to produce electricity efficiently, and then hours again to cool down—so plant operators try to use coal plants as always-on baseload power to decrease wasted energy during the start-up and shutdown periods.)

Geothermal energy may also be used to regulate temperature in buildings, providing an alternative to conventional heating and air conditioning. Hot water from a geothermal source can be pumped directly into buildings for heat. Alternatively, geothermal pumps can utilize the constant temperatures (between 50 and 60 degrees Fahrenheit) found only a few feet underground to cool buildings in the summer and heat them in the winter.

The next generation of geothermal energy, "enhanced" geothermal, aims to harness underground heat sources that otherwise lack water or permeability but are broadly distributed geographically. There is considerable interest in and hope for enhanced geothermal, but the technology is still in development and, like other emerging energy resources, would take time and investment capital to grow to any significant scale.

Key Limiting Factor *Very small sector, would take decades to develop to significant scale.*

Net Energy Ratio

0   10:1   20:1

LIQUID BIOFUELS

Liquid biofuels, and ethanol in particular, have been touted as the solution to U.S. dependence on foreign oil. But most biofuels actually require almost as much energy to produce as they provide. Despite years of development, biofuels remain uncompetitive with fossil fuels.

Biofuels are derived from plant material and fall mainly into two categories: ethanol and biodiesel. In the United States most ethanol comes from corn, but globally it is produced from a variety of plants, including corn, sorghum, sugar, sugar beets, and switchgrass. In a simple chemical process, biodiesel is made from vegetable oil.

The United States currently produces roughly 13 billion gallons (300 million barrels) of ethanol a year, almost entirely from corn—nearly a tenfold increase in over a decade. The ethanol industry has benefited from both an import tariff of 54 cents per gallon on foreign-produced ethanol as well as a subsidy of 45 cents per gallon, costing U.S. taxpayers billions of dollars. The ethanol industry also benefits from laws mandating the blending of ethanol with gasoline.

Unfortunately, producing ethanol is at best a poor use of resources, and at worst a net energy loser. The energy content of ethanol is about two-thirds that of gasoline. An analysis by the think tank Environmental Working Group indicates that blending 10 percent ethanol with 90 percent gasoline (the ratio mandated by the renewable fuel standard) reduces the miles per gallon achieved by almost 4 percent on average. From an energy standpoint, this means that the 10.6 billion gallons of ethanol produced in 2009 in the United States replaced the equivalent of only 7.1 billion gallons of gasoline.

The net energy ratio (energy return on energy invested, or EROEI) for biofuels in general, and corn ethanol in particular, is abysmal. Various studies have estimated the EROEI of corn ethanol at between 0.8:1 and 1.3:1, meaning that we get between 0.8 and 1.3 joules of energy from ethanol for every joule of energy invested in producing that ethanol. The EROEI of gasoline, by comparison, is between 5:1 and 30:1, depending in part on the source of the petroleum.

Additionally, in recent years the ethanol industry's huge purchases of corn as a feedstock for fuel production have caused corn prices to increase, raising the cost of basic food items for the global poor. In response, many ethanol advocates are optimistic about cellulosic ethanol (in particular, switchgrass), since it supposedly would not compete directly with food crops. But cellulosic ethanol also has low net energy, and carries the potential for increased competition for food-growing land.

The EROEI of biodiesel is only somewhat better than that for ethanol. While biodiesel produces fewer emissions (except for nitrogen oxides) than petroleum diesel, its production at industrial scales would inevitably mean further increased competition for arable land and possibly for certain food crops, such as soybeans.

Key Limiting Factors *Extremely low net energy ratio; competes directly with food production for land and feedstock.*

Net Energy Ratio

0   5:1

BIOMASS ELECTRICITY

While small scale biomass heating and cogeneration plants may be a legitimate advance toward a renewable energy economy, large-scale biomass electricity presents the Faustian choice of burning the forest to keep the lights on.

Electricity from biomass is increasingly promoted as a "green" alternative to fossil fuels. As in a coal- or natural gas-burning power plant, biomass fuels are burned to make steam, which drives a turbine to generate electricity. Although biomass can refer to many different potential fuels including crop residue, construction waste, and garbage, the majority of existing biomass-fueled power plants burn wood. As of 2012, hundreds of new biomass-fueled facilities are proposed or under construction around the United States.

The U.S. Energy Information Agency's 2009 electricity generation data shows about 1 percent coming from biomass. Wood has a much lower energy density than fossil fuels, which means that the mass of raw material input per electrical energy output is much higher for biomass than for either coal or natural gas. To meet even a modest percentage of current U.S. electricity demand with biomass would require dramatically increased logging of the nation's forests, and increased removal of woody debris, which is vital for wildlife and healthy forest soils. Industrial biomass energy production, particularly whole-tree harvesting for wood chip–burning power plants, is a growing threat to forest ecosystems.

Biomass burning also produces dangerous air pollution, which is why many physician and medical groups are opposing biomass energy projects. Although biomass energy in theory has no net contribution to global greenhouse gas emissions because the carbon dioxide released during combustion will be recaptured by future forest growth (some question this assumption because climate change may reduce overall forest cover), there is a timing issue that is often overlooked by biomass proponents. The important time horizon for greenhouse gas reductions is the next fifty years. While CO_2 emitted by burning wood will eventually be sequestered, full recovery can be on the order of several centuries. Thus burning wood today may exacerbate global warming in the near term, especially since more wood must be burned compared to other fuels to get the same amount of energy.

Key Limiting Factor *Large-scale development would put pressure on forests and agricultural land.*

Net Energy Ratio

0   10:1   20:1

INDUSTRIAL WIND

Wind power is one of the most successful renewable energy resources, but it does require backup systems to keep generating energy when the wind is not blowing. Additionally, industrial wind developments can have considerable local aesthetic impacts.

Wind power has been utilized by societies for millennia for a variety of functions, including sea transport and milling. Wind turbines today can be small, powering single homes or businesses, or large enough to power a thousand homes. The average industrial wind turbine today stretches roughly 20 stories into the air with a blade diameter of 200 feet and produces enough power for a couple of hundred homes (approximately one or two megawatts of energy).

More than 80 countries around the world have some sort of modern wind power, totaling almost 200 gigawatts of installed capacity. This installed capacity equates to just over 2 percent of annual global electricity consumption. The United States was recently surpassed by China as the world's largest wind power producer, with a total of 44 gigawatts. Denmark, followed by Portugal, and then Spain have the highest proportion of electricity generation from wind, at 21, 18, and 16 percent, respectively. By comparison, the 40 gigawatts produced from wind in the United States represent only 2 percent of total electricity consumption.

The energy return on energy invested for wind power is upwards of 20:1–30:1, which is comparable to that of fossil fuels, and higher than most other renewable resources. However, this figure does not reflect that wind is an intermittent source of energy. Achieving the full benefits of wind power at a large scale requires solving the problem of intermittency with better energy storage technology and smooth integration of baseload generating sources and renewables. Numerous efforts in these areas are underway, including development of smarter electrical grids that may accommodate a high percentage of renewably generated power, but these infrastructure improvements will be expensive.

Although concerns about the negative effects of wind turbines on birds have largely been resolved, other nonenergy-related complications remain, namely local complaints about noise and shadow flicker from blades, and concerns about the visual impact of large facilities. Additionally, wind power tends to be best on mountaintops or offshore—areas that can be tough to reach and may lack electrical infrastructure. New transmission capacity can fragment wildlife habitat. Lastly, due to the fact that wind power is not energy dense, the footprint for a system of wind turbines compared to that of a coal mine or oil and gas field is much larger per energy unit, which may cause increased land-cover degradation and habitat destruction.

Key Limiting Factors *Intermittent, requires backup energy source; large land footprint.*

Net Energy Ratio

| 0 | 10:1 | 20:1 | 30:1 | 40:1 |

SOLAR PHOTOVOLTAIC

The Sun delivers enough energy to the Earth every day to power global society many times over. Solar energy's potential is enticing; photovoltaic technology is improving and the cost is falling. Intermittency, lower energy return on energy investment, institutional barriers, and dependence on rare metals for manufacturing are challenges to solar photovoltaic (PV) gaining a significant share of the global energy portfolio.

Solar PV panels use the energy from the Sun to "excite" electrons into a high energy state, at which point they are converted into electricity. Most photovoltaic panels use crystalline silicon as a base material, but recent advances have led to the use of more scarce elements such as cadmium, tellurium, indium, and gallium. "Thin film" PV panels have also been developed that use less silicon than traditional PV panels. Total global installation of solar PV was roughly 40 gigawatts in 2010, distributed in more than 100 countries. The rapid expansion of manufacturing capacity, particularly in China, has caused solar PV panel prices to drop dramatically, and maturation of the industry is projected to similarly reduce "balance of system" (design, installation, etc.) costs in coming years.

Solar PV offers numerous advantages over fossil fuels for generating electricity. Greenhouse gas emissions are considerably lower over the life of the panel, even when accounting for emissions during construction. Additionally, solar energy is distributed (albeit not evenly) throughout the world, which means many remote populations can produce electricity without constructing inefficient, expensive, and habitat-disrupting long-distance-transmission infrastructure.

Like wind, however, solar energy is intermittent. Not only are there diurnal fluctuations in solar energy but cloud cover, fog, seasonal light availability, and even dust on the panels can severely affect photovoltaic electricity generation. The conversion efficiency (i.e., converting incident solar radiation into electricity) of PV panels is quite low as well, around 15 percent, although estimates vary widely and new technology is incrementally increasing efficiency. The conversion of coal to electricity, by comparison, is over two times more efficient than solar panels. New thin-film PV has been integrated into building facades and roofing, expanding the possibilities of where solar systems can be installed, although it currently has lower conversion efficiencies than conventional PV.

The countries with the fastest growth rates in solar installation are also those with the most aggressive subsidy programs. In Germany, for example, the government for a time was paying more than 60 cents per kilowatt-hour for power from small solar PV systems, which is almost ten times higher than the price of electricity in some parts of the United States. Spain had a similar program that boosted solar electricity generation there. Federal and state incentives, as well as innovative financing programs, have helped stimulate the growth in U.S. solar PV installations, and recent declines in the price of PV panels have prompted some proposed utility-scale solar thermal generating stations to switch to PV.

Key Limiting Factors *Intermittent, requires backup energy source; scalability may be constrained by dependence on scarce or expensive natural resources.*

Net Energy Ratio

| 0 | 10:1 | 20:1 |

CONCENTRATED SOLAR THERMAL

Focusing the relatively dispersed energy from the Sun to produce electricity from steam is a high-tech way of capturing solar energy. Unfortunately, the places where concentrated solar technology works best—deserts—are the same places where a critical component, water, is limited and where impacts on wildlife habitat, including for endangered species, is sometimes inevitable.

Concentrated solar power (CSP) is different from PV systems in that it uses a series of mirrors to focus the Sun's energy into one location where the heat is collected to make steam. Concentrated solar systems therefore produce electricity using the same mechanics as fossil fuels: Steam drives a turbine, which generates electricity. The most popular setup for CSP is called the "parabolic trough" system, which consists of long U-shaped mirrors that reflect sunlight onto a tube positioned above the array. The fluid (generally a synthetic oil) flowing through this tube is heated and is then used to turn water into steam. Concentrated solar power accounts for roughly one gigawatt of global electricity production, with much of the installed capacity located in Spain and the United States.

In CSP, the electricity generation process itself has zero emissions. There are emissions associated with the construction, maintenance, and decommissioning of the facility, but they pale in comparison to those from an average coal- or other fossil fuel-burning plant. But concentrated solar facilities do have a significant physical footprint (like any power plant) and require adequate transmission infrastructure to get electricity to consumers. Conservationists have opposed some CSP plants proposed to be built on U.S. public lands where their construction would negatively affect fragile desert habitat or endangered species. Proper siting on industrial brownfields near existing transmission lines would eliminate these negative impacts of CSP development.

Concentrated solar shares some of the shortcomings of solar PV-generated power. Since both rely on sunlight, they are intermittent sources of energy, which generally means that either natural gas or hydroelectricity must be used as a backup to offset the rapid fluctuations in power output from solar facilities. Additionally, cooling the steam produced at CSP facilities requires massive amounts of water, which is a scarce resource in the sunny, desert environments where CSP facilities are most efficient. On average, CSP plants consume as much water per megawatt of electricity generated as coal plants.

Key Limiting Factors *Intermittent energy source; heavy water user.*

Net Energy Ratio

| 0 | 10:1 | 20:1 |

HYDROGEN

A future hydrogen economy may be technologically possible but is unlikely to be developed on a global scale because of its inherent inefficiencies and capital costs. Hydrogen use may become widespread in some countries, however, and excel for limited uses.

Hydrogen is not, strictly speaking, a primary energy source like coal or oil, since there are no hydrogen reserves to drill or mine. Thus any energy system involving hydrogen will have the added cost of first forming the hydrogen.

On Earth, hydrogen is found only in combination with other elements. The familiar H_2O water molecule, for example, has two hydrogen atoms bound to a single oxygen atom. To acquire hydrogen in a useable form, it has to be split from other substances. The most common method is to split hydrogen off of the methane molecule, CH_4. The vast majority of hydrogen currently produced in the United States comes from a process known as steam reforming, in which steam is reacted with methane at high temperatures and in the presence of a catalyst, releasing carbon dioxide and hydrogen. Another method is electrolysis—ideally using electricity from a renewable source—which strips hydrogen from oxygen in water molecules.

Hydrogen can be burned to power machines such as cars and trucks, to heat homes, or to generate electricity in fuel cells. Its only waste product is water, formed by the reaction with oxygen. Hydrogen fuel cells can be either large centralized facilities or small enough to power a single home. It is a proven, workable fuel: Liquid hydrogen boosts the space shuttle into orbit and hydrogen fuel cells power its electrical systems. A hydrogen economy, however, would be difficult to scale up globally. Fuel cells currently are expensive to build, though once in place, they can provide greenhouse gas–free electricity, especially if the initial electricity used in electrolysis is derived from a renewable energy source such as solar or wind.

Currently, hydrogen use is very modest, but interest in hydrogen is growing because there are no greenhouse gas emissions from burning hydrogen (although greenhouse gas pollution may result from hydrogen production), and the only "waste" product is water. The main barriers to expanded hydrogen use are the huge capital outlays required to develop a national-scale hydrogen production and distribution system, and the low energy return on energy invested.

Key Limiting Factor *Massive investment needed to create hydrogen-related infrastructure.*

MICROPOWER

During the past half century, the energy economy in the developed world has emphasized size; big dams and large, centralized generating stations burning fossil fuels or splitting atoms to generate massive quantities of electricity that is distributed regionally by the grid. Now that trend seems to be reversing.

Small-scale distributed generation, or "micropower," has come to be defined as the growing sector of electrical supply that encompasses combined heat and generation facilities (whether biomass or fossil fueled) plus renewables, excluding large hydro. In 2008 micropower produced 17 percent of the world's electricity, surpassing the global output of nuclear power plants by several percentage points.

Micropower harnesses the most appropriate local energy resources for local use. In practice this may mean solar PV arrays in sunny areas, wind turbines in windy areas, combined heat and power facilities burning crop residue to run a factory in India, or micro-hydro dams in Patagonia. The overarching goals are to democratize power production, improve dependability of the grid, rapidly deploy renewables, and lower costs and emissions by producing electricity near where it is used, thereby eliminating the line losses inherent to long distance transmission.

Micropower generating capacity is usually connected to the grid both to sell excess electricity and to ensure uninterrupted electricity when local generation isn't possible. Even if based on fossil feedstocks such as natural gas, the "radical efficiency" of combined heat and power stations, according to micropower boosters at Rocky Mountain Institute, "typically save at least half—often two-thirds or more—of the fuel, emissions, and cost of making electricity and heat separately." Further greenhouse gas emissions reductions are possible with renewables.

Micropower is the heart of a future distributed power system in which producers of different scales—homeowners, voluntary associations, businesses, schools, or municipalities—generate electricity for their needs and sell the excess back into the grid. This approach has numerous benefits compared to centralized generation, where one large facility produces power and distributes it to an entire region.

Unfortunately, the current grid is ill-equipped to handle a large share of distributed generation based on intermittent renewables such as solar and wind, but efforts to modernize the grid are under way. Various efforts to establish "microgrids" are making progress as well, with notable examples at the University of California at San Diego and on U.S. military bases.

Key Limiting Factor *Economics increasingly favors micropower but institutional barriers and resistance to distributed generation remain in some energy markets.*

REFINERIES

Converting crude oil into its various derivatives (gasoline, diesel, jet fuel, etc.) is no easy task. Refineries have some of the highest rates of greenhouse gas emissions in all of industry, they operate continuously, and, with all of the volatile fuels passing through them, they have a history of dangerous fires and explosions.

Refineries are responsible for turning the various forms of crude oil extracted from underground reservoirs into usable petroleum products, from familiar energy-dense fuels such as gasoline and heating oil to waxes and lubricants. Most of these products are created through a process called fractional distillation, which separates hydrocarbons with different boiling points. The total amount of equipment necessary to refine the petroleum consumed every day is massive, leading to refineries that appear more like small cities. The United States has well over 100 refineries with a total capacity of nearly 18 million barrels per day, the highest in the world. The single largest refinery on Earth is in India, with a capacity of over 1 million barrels per day.

Refining oil requires an immense amount of energy. Refineries power their machinery and processes almost exclusively with oil and natural gas, contributing significantly to greenhouse gas emissions. According to the U.S. EPA, carbon dioxide emissions from on-site energy consumption at refineries are responsible for upwards of 10 percent of all emissions from U.S. industry. Refineries also generate air pollution that can threaten nearby communities. EPA documents have noted that "the petroleum refining industry is far above average in its pollutant releases and transfers per facility" and these chemicals include "benzene, toluene, ethylbenzene, xylene, cyclohexane, 1,2,4-trimethyl-benzene and ethylbenzene," which can be harmful to human health.

In addition to threats to human health caused by pollutants from refineries, the combination of flammable substances, numerous chemical reactions, and high temperatures at refineries leads invariably to accidents. In 2010, four employees were killed in a refinery fire in Anacortes, Washington, and in 2005 a refinery explosion in Texas City killed 15 workers and injured more than 100 others. These are just two of the more egregious recent accidents, and a full history would highlight myriad violations, accidents, and unfortunate deaths.

PIPELINES AND TRANSPORT

The globalized transport network for moving oil, gas, coal, and other fuels is staggeringly large and complex. Every day countless trains, ships, and tanker trucks deliver the fuel that keeps the world economy humming. This transport infrastructure, including hundreds of thousands of miles of pipeline, is vulnerable to accidents and terrorism and is costly in money, energy to maintain, and greenhouse gas emissions.

After the discovery of oil, pipeline transport was quickly adopted as the cheapest delivery method. Pipelines are primarily made of steel, with diameters ranging from a few inches to a few feet, and they are often buried at depths between three and six feet. Oil is pushed through the pipelines by pumping stations—and natural gas by compressor stations—scattered along the route. For natural gas, the United States has more than 300,000 miles of pipeline, 1,400 compressor stations, 11,000 delivery points, 24 hubs, and 400 underground storage facilities. For oil, there are tens of thousands of miles of additional pipeline.

Pipelines are generally considered the safest transport method, although accidents do occur. In 2010, while most public attention was diverted to the Deepwater Horizon oil spill in the Gulf of Mexico, a pipeline in Michigan leaked 800,000 barrels of oil into the local river system. In 2010, a natural gas pipeline exploded in California, killing eight people and creating a crater more than 40 feet deep. However, the biggest regular environmental impact from pipelines is habitat fragmentation: Although much of the pipeline infrastructure is buried, the land cover must remain clear to avoid root obstruction.

Where pipelines are impractical, energy resources are transported by ship, train, or truck. As of 2009, more than one-third of all major shipping vessels in the world were moving oil. Bulk carriers, responsible for coal, grain, iron ore, and other commodities, make up another third. Liquefied natural gas (LNG) vessels account for only 3 percent of all vessels but grew at an annual rate of 11 percent from 2009 to 2010.

Building ships, trains, and trucks requires immense amounts of steel; steel production is a significant source of greenhouse gas emissions. There is also a long history of spills associated with transport, especially oil tankers, which can severely affect regional environments. One can readily find oil-soaked sand on the beaches of Prince William Sound more than two decades after the Exxon Valdez spill. The long-term impacts of the Deepwater Horizon spill are yet to be determined.

The contribution of the energy transport system to climate change is difficult to calculate but as the world shifts from high-energy-content fossil fuels to lower quality forms like tar sands and subbituminous coal, the volume of fuel transported will need to increase proportionally. This will require even more pipelines, ships, trains, and trucks, more fuel to construct and operate them, and will result in more greenhouse gas pollution from the energy sector.

POWER LINES

Power lines serve for electricity the same function that pipelines serve for oil and natural gas. They often produce similar ecological impacts, including habitat fragmentation, and are an aesthetic blight on landscapes. The expanding network of transmission lines has resulted in linear clearcuts through ecosystems around the globe.

Electricity has two drawbacks that oil, natural gas, and coal do not have. It does not exist in nature in a way that humans can harvest directly (we must convert other energy into electricity), and it cannot be stored easily. Yet it is electricity—providing power to illuminate the night and run myriad machines from cell phones to computers—that we most equate with modern society. Electricity consumption tends to grow steadily in developing economies, even while the underlying sources of that electricity (i.e., coal, nuclear, and natural gas) may shift over time. Power lines play the crucial role of transporting the electricity from the point of production to the point of consumption.

Power lines are typically categorized in two groups. *Transmission lines* are high voltage lines used to carry electric current from generating stations to consumption hubs. From hubs, where the current is downgraded to house current, *distribution lines* deliver electricity to the point of consumption.

Power lines can have the same fragmenting effects on wildlife habitat as pipelines. High voltage power lines are allotted a 120-foot right of way (60 feet on each side of the transmission tower) to ensure that the lines are unobstructed from vegetation. This allows companies to clear-cut all natural vegetation within that distance. Clear-cutting forests and other vegetation for pipelines and to accommodate power distribution networks has fragmented forest ecosystems around the world, with substantial impacts on ecosystem integrity. The variety of "edge effects" from such fragemention, particularly the invasion of exotics or weedy species and loss of interior forest habitat, is well described in the scientific literature.

The aesthetic impacts of power lines are more difficult to quantify than ecological costs but are very real to affected communities. New transmission capacity is expensive to build and often highly controversial. There are numerous current campaigns under way fighting proposed power lines, from the "Northern Pass" project in New Hampshire that would bring additional HydroQuebec-generated electricity to the U.S. energy market, to the coalition of activists working to stop a new, roughly 1,200-mile transmission line through southern Chile. That project, proposed in conjunction with a scheme to build multiple large dams on wild rivers in Patagonia, would bisect numerous national parks and national reserves to supply power to urban areas in central Chile.

EMERGING ENERGY TECHNOLOGIES

Virtually every day corporate and university press releases tout the latest technological breakthroughs that will revolutionize the energy sector. Ultimately, some of these innovations will find niche markets, but they generally lack one or more crucial characteristics that make fossil fuels so addictive to a growth-obsessed society.

Thanks to rising oil prices and growing concern about climate change, myriad new energy technologies have emerged in recent years; many are hyped as "game-changing" alternatives to fossil fuels. Freeing society from fossil fuel dependence is undoubtedly a crucial objective, but no single new technology or incremental improvement in existing technology is likely to be the silver bullet that cornucopians expect the market to produce.

Most new energy technologies have significant technical challenges that keep their net energy ratio (energy return on energy invested) relatively low. Wave and tidal power schemes need to operate in corrosive salt water over vast areas under extreme conditions. Algal biofuel needs just the right mix of sun, water, and nutrients and may be difficult to produce at industrial scales. Next-generation solar and wind power relies on scarce or constrained resources like tellurium, gallium, and indium. Fusion power seems perpetually only twenty years from being commercially feasible.

Optimists argue that given sufficient research and development, new energy technologies will evolve, economies of scale will be realized, and costs will be reduced. Very likely this is true, to some degree. Technology improvements will certainly happen. And, much like wind and solar power, there will be specific markets in which these technologies will be useful and possibly even come to dominate. But the fact is, fossil fuels have superlative energy density, versatility, and high net energy (the early conventional oil and coal industries, for example, realized EROEIs of 50:1 or even 100:1). Moreover, our massive globalized economy perches atop a century's worth of physical infrastructure that was built to run on fossil fuels. Emerging energy technologies generally fail in one or more of the crucial categories in which fossil fuels excel: energy density, accessibility, transportability, storability, and sheer abundance.

So while tomorrow's technologies may reduce the toxic effects of the current energy economy, there is no miracle cure for a system that needs structural reform. Perhaps the most worrisome aspect of emerging technologies is the hope they instill in us that technology can ultimately defeat all environmental limits, allowing economic and population growth to continue exponentially, indefinitely. In a finite world, that is a false hope.

Part Four

FALSE SOLUTIONS

INTRODUCTION

False Solutions to the Energy Challenge

If becoming energy literate and understanding the philosophical foundations of the energy/growth nexus is a necessary first step toward useful participation in the societal conversation about energy, then what is the second step? There may be multiple correct answers to that question, but certainly one possibility is to study the proposals of those who avoid discussion of fundamental restructuring and instead offer "solutions" that merely tinker around the edges of the current system.

By dissecting the arguments advanced by boosters of various energy sources and policies, one gains skill in recognizing ideas designed to serve the particular agenda (usually profit-oriented) of a narrow constituency. Such arguments are often framed in the familiar rhetoric of "keeping inexpensive energy flowing to create jobs and grow the economy." Anyone who has developed a systemic critique of the current energy economy will readily see through that boilerplate and have thoughtful opinions on why scale matters, why aesthetics are important, and why efforts to solve problems caused by growth with ever-more growth are dubious. These opinions will inform one's judgment about whether a proposal offers legitimate progress toward a nature-friendly energy system or is a false solution that continues, or worsens, the status quo. Energy policies that emphasize gigantism and endless economic growth will be immediately suspect.

To be fair, deeming something a "false" solution leaves little room for nuance that the real world demands. Energy policy, like life, is not always neatly bifurcated into easy categories of yes or no, true or false. Can America drill its way out of reliance on "foreign" oil? No. But increased domestic drilling (and an economic recession) has lessened the percentage of oil we import, and there are compelling—albeit difficult to implement—blueprints for how to wean ourselves off oil entirely. Do megadams kill wild rivers? Yes. But developed countries with a high percentage of hydropower, such as Norway, have far lower carbon foot-prints from the electrical sector of their energy portfolio.

The energy "solutions" considered in this section are not comprehensive. But the ideas considered—"drill!," "frack those wells!," "build more nukes and dams!," "coal is clean!," "burn biomass!," "unconventional hydrocarbons will save us!," and "don't worry because the government will adequately regulate all of it!"—are ubiquitous. These and similar viewpoints are well articulated by politicians and industry groups. Open a newspaper or weekly news magazine and there will be flashy advertisements singing the praises of "clean" coal, celebrating the can-do spirit of the natural gas industry, or describing how oil company X has the greenest agenda since John Muir.

That's corporate spin, of course, but such advertising accurately reflects the desires of a society built upon near-religious devotion to the idea of economic growth and cultural progress through technological innovation. Perhaps the ultimate techno-optimist idea, however, is the notion of geoengineering—that we can keep on running an energy economy that cooks the planet until the heat becomes intolerable, and then we'll simply reengineer the global climate. It may well be the ultimate false solution to the energy challenges we face.

DRILL BABY DRILL

Why It Won't Work for Long-Term Energy Sustainability

DAVID HUGHES

GLOBAL FOSSIL FUEL *consumption has accelerated rapidly over the last few decades, requiring an enormous stream of resources to meet even current demand. Claims that future demand can be met simply by opening new areas to exploration and increasing production of relatively new resources like deepwater oil, tar sands, shale oil, oil shale, and shale gas ignore both the scale of global fossil fuel consumption and the technical challenges of producing such resources at a sufficient rate.*

❖

"Drill Baby Drill" was famously uttered by Republican Sarah Palin during a 2008 vice-presidential debate as a response to America's addiction to imported oil.[1] In mid-2011, Republican presidential candidate Michelle Bachmann claimed to be able to— if elected— reduce the price of gasoline to less than $2 per gallon from its then-current price of nearly $4.[2] These politicians, despite their naiveté on matters geological, understand the correlation between energy consumption and the American Dream aspired to by the electorate. Given that fossil fuels provide the lion's share of the world's energy at present, and that oil—the largest source of energy in the world—has been at historically high price levels recently, what is the outlook for a supply-side solution to America's energy dilemma?

THE CONSUMPTION SPIRAL

Fossil fuels represent an incredibly dense and convenient form of energy— the legacy of hundreds of millions of years of fossilized sunshine preserved by generally very inefficient processes. Oil, for example, accumulated over a period of 500 million years. If we assume that 3 trillion barrels of oil will eventually be recovered and burned (the most common estimate), this means roughly fourteen and a half thousand years' worth of preserved fossilized sunshine is consumed *each day* at the current global consumption rate of 87 million barrels per day (mbd).

Prior to 1850 more than 80 percent of energy consumption was provided by renewable forms of energy, mostly biomass, with the balance provided by coal. Today, 84 percent of the average world citizen's energy is provided by fossil fuels, with most of the balance provided by nuclear and large hydropower. Fossil fuels have allowed the per capita consumption of the average world citizen to increase nearly ninefold since 1850. Increased energy availability for the production of food and other commodities has also allowed global population to increase nearly sixfold over this period. As a result, the world is consuming 49 times as much energy as in 1850, 89 percent of which is nonrenewable (oil, gas, coal, and uranium).

The pace and scale of this growth are astounding. Fully 90 percent of all fossil fuels have been consumed since 1937. *Half* have been consumed since 1985. Conventional wisdom is that coal has been replaced to a large degree by oil and natural gas. This is not true: The average world citizen consumes the same amount of coal per capita today as in 1910, and 90 percent of all coal ever consumed has been burned since 1911. The consumption of oil merely added to per capita energy consumption, as opposed to displacing coal. Natural gas and nuclear energy added still further to energy throughput. Ninety percent of all oil consumed by humankind has been burned since 1961, and 90 percent of all natural gas since 1966. The rate of consumption has accelerated rapidly over the last couple of decades: 50 percent of all oil since 1988, 50 percent of natural gas since 1992, and 50 percent of coal since 1975.

Increased consumption of energy is highly correlated with rising Gross Domestic Product (GDP) and the so-called good life enjoyed by the industrialized nations of the world—represented by the Organization for Economic Cooperation and Development (OECD). The OECD countries make up 18 percent of the world's population and consumed four times as much energy per capita as the non-OECD world in 2010. The United States, with 4.5 percent of the world's population, consumed 19 percent of the world's energy in 2010—on a per capita basis this is considerably more than the OECD average, four times that of China, 17 times that of India, and 49 times that of Bangladesh.

A major problem going forward is that the developing world aspires to the developed world's levels of energy consumption. China, for example, has increased its total energy consumption by 134 percent over the past decade, and is now the largest energy consumer in the world. Yet China consumes only one-quarter as much per capita as the United States. This represents a profound geopolitical conundrum for the future given the desire for growth in energy consumption in the developing world, the desire to maintain consumption and GDP growth in the industrialized world, and physical limits to the ability to grow or even maintain global energy supplies.

If we look at oil—the premium fuel for transportation and the only fossil fuel that is largely globalized in terms of price—the stark inequities of consumption between the developed and the developing world become even more evident. OECD nations consume five times as much oil per capita as non-OECD nations. The United States in 2010 consumed nine times as much oil per capita as China and 22 times as much as India. China's oil consumption has nearly doubled in the last decade, yet it consumed less than half of the oil consumed by the United States in 2010. India's oil consumption, which has increased by 50 percent over the last decade, is still less than a fifth of the total oil consumed by the United States. Yet all three countries are heavily dependent on imports—the United States 58 percent, China 55 percent, and India 75 percent—and all are competing in the world market for oil imports.

Global oil production has been on an undulating production plateau since early 2005, despite historically high prices. In 2010, 79 percent of global production came from countries that are past their peak levels of production. Although some of these countries (such as Saudi Arabia) have significant surplus capacity and may exceed their previous production highs in the future, most countries that are significantly below their peak production levels are probably permanently past peak. The global peak of oil production is likely to occur by 2020—and possibly much sooner.

THE FOSSIL FUEL ENERGY SUPPLY DILEMMA

Economics dictate that the low-hanging energy fruit gets picked first: the large, easy to find, light oil fields; the highly productive conventional gas fields; and the thick near-surface coal seams. As these are depleted, attention turns to more difficult and higher cost deposits: deepwater oil, polar oil, tar sands, and shale oil; tight gas, shale gas, and coalbed methane; and thinner, deeper, and lower quality coal seams. The easy fossil fuel energy in the world has largely been identified and exploitation is under way or complete, and we now increasingly rely on more difficult and higher cost sources.

If one wants to count up all of the hydrocarbon molecules in the ground, the only ones that count are those that can be profitably extracted. Technology and price can allow increased access to resources as technologies get better and/or prices go higher. The ultimate limiting barrier to access, however, is net energy, defined as the surplus energy available from a resource after subtracting the energy inputs required to explore for and produce it. Fossil fuel resources with a net energy of zero or less make no economic sense to recover, and unfortunately this applies to the bulk of the *in situ* resources remaining today.

These concepts are mainly lost on politicians, many economists, and most industry promoters. We are assured that there is one hundred years' worth of accessible natural gas in the United States thanks to shale gas; more than a trillion barrels of oil shale in Colorado, Utah, and Wyoming; abundant shale oil in North Dakota and Texas; and that if all else fails we can still liquefy coal and gas to produce synthetic oil.

A useful concept to understand the issues in maintaining flow rates adequate to meet future demand for fossil fuels is that of the "tank" and the "tap." The tank is the size of the in situ resource, and the tap is the rate at which it can be produced, given capital investments, infrastructure, time, and operating costs. For example, the high-porosity structural trap that defines Ghawar—the largest oil field in the world, which was discovered in 1948 in Saudi Arabia—had both a very large tank and a large tap. That highly pressured reservoir produced prolifically from low-tech vertical wells, with a very high net energy profit: 100:1 or more. Nowadays, Ghawar is getting long in the tooth, and the latest technology must be applied to keep the tap open—yet it still produces half of the Saudis' oil.

The elephants like Ghawar have been found and the easy onshore discoveries made. New discoveries are in environments that are much more hostile, resulting in much higher costs: deepwater offshore, subsalt deposits in the deep Atlantic, the polar icecaps and so forth. BP's deepwater Macondo well which blew out in the Gulf of Mexico in 2010, sinking the Deepwater Horizon rig, would have cost over $100 million if it had been completed successfully; as it was, it ending up costing many billions of dollars and the costs are mounting. Several deepwater dry wells off the coast of Nova Scotia have cost more than $100 million each, and wells in the subsalt Santos Basin off the coast of Brazil are likely to make those figures seem like a bargain. It takes a lot more money, time, infrastructure, and effort to keep the taps open on such new discoveries— with a correspondingly lower net energy yield.

Industry promoters such as the widely quoted Daniel Yergin of IHS CERA[3] tell us that we can recover much more oil from fields already discovered by using high-tech "enhanced" recovery methods. This is true— however, the tap on spent fields using enhanced tertiary recovery typically only opens to a trickle compared to peak production volumes. Indeed, the first oil field in Canada, discovered back in 1858, was still producing even a few years ago—but only a few barrels a week. At the current global oil consumption rate of 31 billion barrels per year, tertiary recovery projects don't have the tap throughput rate to come close to keeping up.

This leaves us looking to purportedly vast "unconventional" oil resources like Canadian tar sands, Venezuela's extra-heavy oil, natural gas liquids, shale oil in formations like the Bakken (North Dakota) and Eagle Ford (Texas), and oil shale in Colorado, Utah, and Wyoming. Politicians point glibly at these sources, divide current annual consumption into the vast purported quantities, and come up with decades or centuries of supply. But the problem with these resources is the size of the tap:

- Tar sands require many years, large capital expenditures, and large environmental impacts to grow production. As a result, they require high prices, just to break even, of some $70–$80 per barrel for new projects. Although the Canadian tar sands have been under development for forty years, the tap is currently only open at 1.5 mbd out of an 87 mbd global oil demand. The most favorable tar sands resources in terms of thickness, depth, and concentration are extracted first. The net energy profit is low, at between 3:1 and 6:1, depending on whether the bitumen is extracted by underground or surface methods, respectively—and most of the purported 1.7 trillion barrels will not be recoverable at a net energy profit. The Venezuelan "extra heavy" oil in the Orinoco Basin faces similar limitations, and in fact, Venezuela's oil production peaked back in 1970.

unlikely to happen due to unfolding physical and economic constraints. We ignore such constraints at our peril.

CONCLUSIONS

The "Drill Baby Drill" mind-set belies any understanding of the degree to which fossil fuels underpin current energy systems; the physical limitations of these resources as they decline in quality and require ever-increasing levels of effort, capital, and environmental risk to maintain production; and the fact that these limitations ultimately spell the end of the continual growth paradigm that has dominated economic thinking since the beginning of the hydrocarbon age. Fossil fuels will of necessity be with us for a long time to come, but it behooves us to recognize their limitations. Adopting strategies to radically reduce consumption while maintaining a reasonable quality of life are mandatory. Energy consumption cannot exceed energy supply, and if we don't proactively manage our energy future, Mother Nature will certainly take care of it for us.

NUCLEAR POWER AND THE EARTH

RICHARD BELL

NUCLEAR POWER HAS *gained renewed traction in recent years as the world's countries seek climate-friendly alternatives to fossil fuels. But public health risks, continued dependence on public subsidies, and the seemingly intractable problem of long term nuclear waste storage make nuclear power a poor solution to our energy needs.*

❖

Nuclear technology is leaving a visible legacy on and within the Earth that could last longer than the human species itself. Giant craters mar the surface of countries that have tested nuclear weapons underground. Leaking high-level nuclear waste tanks at Hanford, Washington, have created a vast basin of radioactive groundwater, some of which has already reached the Columbia River.

Nuclear technology is also creating very real problems for us right now. The 2011 Fukushima disaster reminded the world that every nuclear power plant is a potential public health catastrophe, while the spread of civilian nuclear power heightens the risk of nuclear weapons proliferation. Meanwhile, the U.S. nuclear power industry is still very much dependent on public subsidies, despite more than sixty years of strong government support.

Nevertheless, nuclear power remains a major part of our energy mix, and in recent years it has gained new traction as countries seek climate-friendly alternatives to fossil fuels. This renewed push is not because we are close to solving the problems of nuclear power, but rather because of a deep—and ultimately misguided—faith in the engineers, scientists, and leaders we trust to guarantee the safety of nuclear power plants.

ATOMS FOR WAR AND PEACE

To fully understand the impact of nuclear technology, we must first clear our heads of the fundamental deceit of the nuclear age: namely, that there exists such a thing as "Atoms for Peace." This little phrase, launched by President Eisenhower in a famous speech to the United Nations on December 8, 1953, created a powerful belief that the world's governments were capable of cleanly separating—politically, psychologically, and technologically—the evils of "atoms for war" from the benefits of "atoms for peace."

Unfortunately for humanity, the atoms of uranium and plutonium do not know the difference between war and peace. Nuclear technology is indivisible: Whether you're building bombs to turn cities into wastelands or boiling water to make electricity, the underlying physics and the raw materials are the same. The processes used to slightly "enrich" naturally occurring uranium for reactor fuel can also be used to produce highly enriched uranium for weapons.

When Eisenhower spoke to the U.N., he spoke to people for whom the first images of nuclear technology were the searing photographs of Hiroshima and Nagasaki, blasted landscapes where nothing was left standing except the skeletons of a few steel structures. (Later came the images of whole islands obliterated by testing in the Pacific, and mushroom clouds, heaving and boiling, carrying radioactive fallout into the stratosphere to be distributed by the winds across the entire planet). Eisenhower began his speech by reviewing the "dark chamber of horrors" of the arms race but soon turned to his vision of a world in which nuclear technology would be put to use for peaceful purposes instead:

> *The United States knows that if the fearful trend of atomic military build-up can be reversed, this greatest of destructive forces can be developed into a great boon, for the benefit of all mankind. The United States knows that peaceful power from atomic energy is no dream of the future. The capability, already proved, is here today.*

Nuclear power plants were supposed to produce electricity "too cheap to meter." They were presented as the apotheosis of twentieth-century technology, alongside equally shiny images from the space program: lots of highly polished gleaming metal, fantastically elaborate control rooms, and workers dressed in surgically white uniforms. In outdoor shots, the plants appeared to be antiseptically clean, with bucolic backgrounds of meadows, green trees, and blue sky flecked with clouds. There was nary a hint of the ferociously dangerous radioactive process going on inside these reactors. The contrast between such images and those from Hiroshima and Nagasaki could not have been greater.

SAFETY AND HEALTH

Given nuclear power's arrival as a child of the nuclear weapons program, it was only natural that the public was concerned about how dangerous these plants were. In 1957, scientists at the Brookhaven National Laboratory issued a report that predicted that an accident at a small nuclear plant could cause 3,400 deaths, 43,000 injuries, and $7 billion in property damages. When an update of this report suggested that an accident might affect an area "equal to that of the state of Pennsylvania," the Atomic Energy Commission suppressed it.

Nuclear proponents have always insisted that reactors are "safe." They avoid even using the word "accident." For example, officials referred to the core meltdown at the Three Mile Island nuclear power plant in 1979 as an "event," an "incident," an "abnormal evolution," or a "normal aberration."

But as we learned after Three Mile Island, the fundamental cause of nuclear accidents is not a technical problem, but a mental one. A presidential commission found that witnesses repeatedly referred to the need to change the "mindset." It concluded that, over the years, "the belief that nuclear power plants are sufficiently safe grew into a conviction. One must recognize this to understand why many key steps that could have prevented the accident at Three Mile Island were not taken."

CHERNOBYL

At the Chernobyl nuclear power plant in Ukraine, Soviet nuclear engineers were so convinced that their design was safe that they did not build a containment building around the reactor to prevent radioactive materials from rapidly escaping into the environment if there were an accident.

In April 1986, the Chernobyl plant experienced a sudden power output surge during a systems test. This was followed by a series of explosions and a fire that sent a plume of highly radioactive smoke into the atmosphere, affecting parts of Europe and the western Soviet Union. The first victims at Chernobyl were the workers and emergency response crews, who absorbed huge doses of radiation; two died within hours, 28 died within three months.

In a clear example of how the nuclear mindset affects our ability to assess the dangers of nuclear power, Soviet authorities had failed to surround the Chernobyl plant with enough sensors to assess radioactive emissions from an accident. So the estimates of how much radiation was released vary widely, resulting in equally wide estimates of the number of accident-related deaths.

In 2005, a joint study by the World Health Organization, the United Nations Development Programme, and the International Atomic Energy Agency settled on an estimate of 4,000 total deaths from radiation exposure. In 2006, Greenpeace released a study with a radically higher estimate of 200,000 additional deaths, including more than 93,000 fatal cancers.

FUKUSHIMA

Despite all the official exhortations, even serious accidents like Three Mile Island and Chernobyl have not been sufficient to change the nuclear mindset, as we have learned from the post-mortems on the latest nuclear disaster at Fukushima on March 11, 2011. TEPCO, the large Japanese utility that owned the six reactors at the ocean-front site, had not learned the very simple lesson from Chernobyl about the need for more sensors to provide post-accident data. Once again, there are huge uncertainties about the amounts of radiation released from the crippled reactors and the spent fuel pools.

The Japanese government and TEPCO initially downplayed the threat, following the pattern of officials at Three Mile Island and Chernobyl. In June, 2011, the government released an estimate of the total amount of biologically dangerous cesium-137 (15,000 terabecquerels) released. But in October, an international team released a new estimate *more than double* the June estimate (35,800 terabecquerels). Some of the data for the October study came, ironically, from the International Monitoring System, a global network of sensors that was put into place to monitor nuclear weapons tests under the Comprehensive Nuclear Test Ban Treaty (which the United States has never ratified).

The October study also cast serious doubt on the government claim that the fire in the spent fuel pool of Reactor 4 did not release significant amounts of radiation. The authors concluded that this spent fuel pool was responsible for the release of large amounts of cesium-137. The level of cesium-137 dropped dramatically once TEPCO flooded the pool with water more than a week after the start of the accident, a step that probably would have greatly reduced the release if the utility had acted earlier.

MISSING PICTURES

Nuclear power plants are the most visible manifestation of nuclear technology. But the plants themselves sit in the middle of a cycle, the beginnings and endings of which are far less well known. On the front end of this cycle, for both nuclear weapons and nuclear plants, are operating mines and thousands of abandoned mines. The public rarely sees images from this part of the process, where millions of tons of radioactive mill tailings accumulate in piles and ponds. These tailings contain 85 percent of the radioactivity of the original ore, and they remain dangerous for thousands of years.

The history of the management of mill tailings is a dismal story. In the rush to mine uranium during the huge buildup of nuclear weapons in the early decades of the Cold War, mining companies simply left the tailings in piles on land, or in poorly constructed ponds. When the demand for uranium slowed in the 1970s, many companies abandoned their mines and mills. The federal government was late to respond. It was not until 1978 that Congress made the Department of Energy responsible for dealing with tailings at 24 milling sites in ten states, as well as at 5,200 other related properties. These included more than 500 abandoned mines on Navajo land, where more than 4 million tons of uranium ore had been mined.

Nor does the public see much of the back end of the production cycle: the highly radioactive spent fuel rods from nuclear power plants, and the high-level liquid wastes from producing plutonium for nuclear weapons. In one sense, the back end of the cycle does not even exist in the United States for the waste from nuclear power plants, since there is no permanent waste storage site— more than fifty years after the first commercial nuclear plant went into operation.

And then there are the nuclear power plants themselves. The intense radioactivity generated during normal operations turns the plant into a behemoth of nuclear waste. In an age of shrinking budgets, there will be great pressure to mothball these reactors and leave them sitting on the land-scape for a century or two while the radioactivity inside slowly dies away.

HIDDEN SUBSIDIES

Comparing the cost of nuclear-generated electricity with that of electric-ity from other sources is difficult because of the large number of assump-tions about what costs to include and what costs to exclude. Prices for coal, for example, usually do not include any costs associated with global warming or increased lung disease, while prices for nuclear power usually do not include costs associated with long-term waste disposal.

Nevertheless, even in a direct comparison considering only construc-tion, operations, and fuel, nuclear power does not fare well. Recent studies from the Massachusetts Institute of Technology and the National Bureau of Economic Research have estimated the cost of electricity from nuclear power to be 30–50 percent higher than from coal or natural gas, largely due to the higher costs for plant construction and the higher risk of default.[1] Hidden government subsidies can introduce large distortions in cost esti-mates—and the nuclear industry has benefited enormously from dozens of direct and indirect subsidies. In 2010, the Union of Concerned Scientists released the most rigorous study to date of government subsidies to the nuclear industry. After adding up subsidies from 1960 to 2008, author Doug Koplow reached a stunning conclusion. These subsidies were:

> ... more valuable than the power produced by nuclear plants over that period. Without these subsidies, the industry would have faced a very dif-ferent market reality—one in which many reactors would never have been built, and utilities that did build reactors would have been forced to charge consumers even higher rates.[2]

Indeed, the subsidies were so high, Koplow concluded, that "buying power on the open market and giving it away for free would have been less costly than subsidizing the construction and operation of nuclear power plants. Subsidies to new reactors are on a similar path."

THE TEN–MILLENNIUM DANGER

Ensuring the safe day-to-day operation of nuclear plants has proved to be extraordinarily difficult. And the damage from the spent fuel pool at

Fukushima's Reactor 4 highlights an even more difficult problem: how to dispose of the highly radioactive wastes from nuclear plants for periods of time far in excess of the ten thousand years of recorded human history.

Nuclear proponents in the United States have recited the same mantra for decades when asked about disposing of nuclear waste: It should go into an underground repository. However, this seemingly sensible declaration is not based on scientific studies showing that identifying and building such a repository is possible. Geologists looking for a permanent underground repository site have repeatedly learned that the Earth's geology is much more complex than anyone imagined. And the political struggles over site selection have become metaphorically radioactive.

The most recent effort to build a high-level waste repository has been at Yucca Mountain in Nevada. The selection of this site and its subsequent history has been a highly politicized process from the very beginning. Nevada's politicians, led by Democratic Senator Harry Reid, have waged a bitter fight to prevent Yucca Mountain from ever opening. In 2009, President Barack Obama officially abandoned Yucca Mountain, announcing the start of yet another process to find another site. Meanwhile the ever-growing accumulation of spent fuel rods stored on-site at nuclear plants increases the dangers of Fukushima-like accidents, or the intentional release of highly radioactive materials from terrorist attacks on the poorly protected spent-fuel storage areas.

IS NUCLEAR POWER WORTH THE CANDLE?

The lure of putting nuclear energy to work is mythological in its intensity. As the Department of Energy puts it:

> *One ton of natural uranium can produce more than 40 million kilowatt-hours of electricity. This is equivalent to burning 16,000 tons of coal or 80,000 barrels of oil.*

Nor is it surprising that, once scientists and engineers had seen the results of unleashing the power of the atom on Hiroshima and Nagasaki, they felt an overwhelming desire to find ways to turn this destructive force to peaceful ends.

At a philosophical level, the problem we face in trying to build "safe" nuclear plants requires us to confront one of humanity's oldest and most well-documented weaknesses: hubris. After several hundred years of ever-more-powerful scientific and technological triumphs, today's leaders find it hard to consider, much less admit, that some tasks may be so difficult that the risks of attempting them are greater than the reaped rewards.

THE FALSE PROMISE OF "CLEAN" COAL

JEFF GOODELL

THE COAL INDUSTRY's *slick advertisements promoting "clean coal" employ twenty-first-century media techniques to keep us locked into a nineteenth-century energy economy. A classic greenwashing campaign, it uses the iconography of sexy technology and down-home Americana to maintain the status quo: Big Coal's influence over energy politics.*

❖

Several years ago, in Gillette, Wyoming, I fell into a long conversation with the vice president of a large American coal company about coal's public image problem. Gillette is in the center of the Powder River Basin, the epicenter of the coal boom in America, where 60-foot seams of coal lay just below the surface. This vice president, who did not want his name to appear in print, was deeply concerned about coal's future and expressed frustration with environmental attacks on coal, suggesting that it was all a problem of perception: "People don't like coal because it's black," he told me. "If it were white, all our problems would be solved."

Whenever one of those slick ads for "clean coal" pops up on CNN, I think about that conversation in Gillette. The 35-million-dollar "clean coal" campaign, spearheaded by a coal industry front group called American Coalition for Clean Coal Electricity (formerly known as Americans for Balanced Energy Choices), is nothing less than a nationwide effort to paint coal white.

And to the coal industry's credit, they're doing a pretty good job. Republicans and Democrats alike tout "clean coal" as the solution to America's energy troubles. The logic is simple. America has lots of coal. We are a technologically advanced society. Ergo, we can clean up coal. What's the problem?

Well, here's one: "Clean coal" is not an actual invention, a physical thing—it is an advertising slogan. Like "fat-free donuts" or "interest-free loans," "clean coal" is a phrase that embodies the faith that there is an easy answer for every hard question in America today. We can have wars without sacrifice. We can borrow more than we can afford without worrying about how we'll pay it back. We can end our dependency on oil by powering our SUVs with ethanol made from corn. And we can keep the lights on without superheating the climate through the magic of "clean coal."

Mining and burning coal remains one of the most destructive things human beings do on this Earth. It destroys mountains, poisons water, pollutes the air, and warms the atmosphere. True, if you look strictly at emissions of smog-producing chemicals like sulfur dioxide, new coal plants are cleaner than the old coal burners of yore. But going from four bottles of whiskey a week down to three does not make you clean and sober.

Of course, the "clean coal" campaign is not about reality—it's about perception. It's an exercise in rebranding. Madison Avenue did it for Harley Davidson motorcycles and Converse shoes. Why not Old King Coal? It's not a difficult trick—just whip out some slick ads with upbeat music and lots of cool twenty-first-century technology like fighter jets and computers. Run the ads long enough, and people will believe.

But the real goal of the campaign is not simply to rebrand coal as a clean and modern fuel—it's to convince energy-illiterate TV viewers that the American way of life depends on coal. The ads remind us (accurately) that half the electricity in America comes from coal, then they show images of little girls getting tucked into bed at night or Little Leaguers playing ball under the lights.

The subtext is not simply that, without the electricity from coal, the lights will go out and your family will be plunged into darkness. It's that, without coal, civilization as we know it will come to an end. As one utility industry executive asked me while I was reporting my book *Big Coal*, "Have you ever been in a blackout? Do you remember how scary it was?"

From the coal industry's point of view, this is a brilliant way to frame the argument. If the choice is between coal or chaos, they win. This framing also disarms environmental arguments—yes, it's too bad that mountaintop-removal mining has destroyed or polluted 1,200 miles of streams in Appalachia and that the Environmental Protection Agency projects a loss of more than 1.4 million acres—an area the size of Delaware—by the end of the decade.

But hey, if it's a choice between flattening West Virginia and keeping our lights on, good-bye West Virginia! That's a false choice, of course.

The coal industry may not want to acknowledge it, but we're living in the twenty-first century now. We have indeed figured out other ways to generate electricity besides burning 30-million-year-old rocks. And with each passing year, those alternatives are getting cheaper and smarter.

Wind is already less expensive than coal in many parts of the country, and so is large-scale solar thermal. Solar PV costs are plunging. Google is exploring enhanced geothermal. The creaky old electricity grid will soon morph into a system that looks more like the Internet, driving big gains in efficiency and allowing for real-time pricing of a kilowatt of power.

This does not mean we can shut down every coal plant tomorrow. But it does mean that coal is no longer the engine of civilized life as it has been since the Industrial Revolution.

Big Coal is best understood as a beast of inertia, pushed along by hundreds of billions of dollars worth of heavy metal infrastructure, and kept on track by an army of lobbyists, and our own ignorance of what goes on behind the light switch.

That may be changing. Even seven-year-olds know that the accumulation of greenhouse gases in the atmosphere, especially carbon dioxide, is warming the planet. Coal is by far the most carbon-intensive of fossil fuels, and currently in the United States there is no financial cost to dumping carbon dioxide into the atmosphere. Big Coal will vigorously fight any legislation that actually puts a serious price on carbon dioxide, because once that market signal is put in place, coal's reign as a "cheap" energy source is officially over.

Big Coal insists they have the solution. It's called "carbon capture and storage." In most scenarios, capturing and storing carbon dioxide from coal involves building a new kind of power plant that uses heat and pressure to gasify the coal, instead of burning it directly. In these new plants, the carbon dioxide can be removed and then injected underground in abandoned gas and oil wells or deep saline aquifers.

Big Coal would like us all to believe that capturing and storing carbon dioxide from these new coal plants is a slam-dunk technology—but one that's not quite ready for prime time yet (capturing these emissions from existing combustion coal plants, while theoretically possible, is far too expensive and ineffecient to be taken seriously by anyone but the most die-hard coal boosters).

Of course, Big Coal has always been better at touting new technology than actually deploying it. Yes, there are serious questions about how much it will cost to build new coal plants that can capture and store carbon dioxide, how soon it will happen, and whether or not the technology can scale up quickly enough to really make a difference. But it's not just technology that's holding back carbon capture and storage. It's politics. Without a price on carbon, there is little incentive to do anything serious about carbon dioxide emissions from coal plants.

Meanwhile, the need to reduce emissions grows more urgent every year. As NASA climatologist James Hansen has repeatedly pointed out, continuing to burn coal the old-fashioned way is a sure-fire way to melt Greenland and turn Miami into an aquarium.

In the end, the "clean coal" campaign is about using the tools of the twenty-first century to keep us locked in the nineteenth century. Like other greenwashing campaigns, it's about using the iconography of sexy technology and down-home Americana to maintain the status quo.

These campaigns always pretend to offer inspiration about what we can do in America if we set our minds and hearts to it, but in fact the real message is what we *can't* do: We can't power America without coal, and we can't pass meaningful carbon legislation without wrecking the American economy.

This is why the false promise of "clean coal" is dangerous. The goal is not to solve our problems, but to perpetuate our addiction. In one ad, the narrator even adopts the feel-good language of substance abuse and recovery: Cleaning up coal is a "big challenge," he explains. "But we've made a commitment—a commitment to clean."

After decades of stoking the engines of denial and obfuscation on global warming, it's nice that Big Coal wants to be a good citizen. But just because your pusher decides to shower and shave, don't delude yourself into thinking that he cares about your welfare.

His real goal is to keep you hooked.

THE WHOLE FRACKING ENCHILADA

SANDRA STEINGRABER

A NATURAL GAS *production boom made possible by new technology—horizontal drilling and hydrofracking—is being touted as a "game changer" in the energy landscape. It also portends massive damage for families, communities, and wildlife as drilling-related pollution violates the bedrock, the atmosphere, and everything in between.*

❖

I have come to believe that extracting natural gas from shale using the newish technique called hydrofracking is *the* environmental issue of our time. And I think you should, too.

Saying so represents two points of departure for me. One: I primarily study toxic chemicals, not energy issues. I have, heretofore, ceded that topic to others.

Two: I'm on record averring that I never tell people what to do. If you are a mother who wants to lead the charge against vinyl shower curtains, then you should. If the most important thing to you is organic golf courses, then they are. So said I.

But high-volume slick water hydrofracturing of shale gas—fracking—is way bigger than PVC and synthetic fertilizer. In fact, it makes them both cheaply available. Fracking is linked to every part of the environmental crisis—from radiation exposure to habitat loss—and contravenes every principle of ecological thinking. It's the tornado on the horizon that is poised to wreck ongoing efforts to create green economies, local agriculture, investments in renewable energy, and the ability to ride your bike along country roads. It's worth setting down your fork, pen, cellular phone—whatever instrument you're holding—and looking out the window.

The environmental crisis can be viewed as a tree with two trunks. One trunk represents what we are doing to the planet through atmospheric accumulation of heat-trapping gases. Follow this trunk along and you find droughts, floods, acidification of oceans, dissolving coral reefs, and species extinctions.

The other trunk represents what we are doing to ourselves and other animals through the chemical adulteration of the planet with inherently toxic synthetic pollutants. Follow this trunk along and you find asthma, infertility, cancer, and male fish in the Potomac River whose testicles have eggs inside them.

At the base of both these trunks is an economic dependency on fossil fuels, primarily coal (plant fossils) and petroleum (animal fossils). When we light them on fire, we threaten the global ecosystem. When we use them as feedstocks for making stuff, we create substances—pesticides, solvents, plastics—that can tinker with our subcellular machinery and the various signaling pathways that make it run.

Natural gas is the Dr. Jekyll and Mr. Hyde of fossil fuels: When burned, natural gas generates only half the greenhouse gases of coal, but when it escapes into the atmosphere as unburned methane, it's one of the most powerful greenhouse gases of them all—over 20 times more powerful than carbon dioxide at trapping heat and with the stamina to persist nine to fifteen years. You can also make petrochemicals from it. Natural gas is the starting point for anhydrous ammonia (synthetic fertilizer) and PVC plastic (those shower curtains).

Until a few years ago, much of the natural gas trapped underground was considered unrecoverable because it is scattered throughout vast sheets of shale, like a fizz of bubbles in a petrified spill of champagne. But that all changed with the rollout of a drilling technique (pioneered by Halliburton) that bores horizontally through the bedrock, blasts it with explosives, and forces into the cracks, under enormous pressure, millions of gallons of water laced with a proprietary mix of poisonous chemicals that further fracture the rock. Up the borehole flows the gas. In 2000, only 1 percent of the natural gas we produced was shale gas. Ten years later, almost 20 percent is.

International investors began viewing shale gas as a paradigm-shifting innovation. Energy companies are now looking at shale plays in Poland and Turkey. Fracking is under way in Canada. But nowhere has the technology been as rapidly deployed as in the United States, where a gas rush is under way. Gas extraction now goes on in 32 states, with half a million new gas wells drilled in the last ten years alone. We are literally shattering the bedrock of our nation and pumping it full of carcinogens in order to bring methane out of the Earth.

And nowhere in the United States is fracking proceeding more manically than Appalachia, which is underlain by the formation called the Marcellus Shale, otherwise referred to by the *Intelligent Investor Report* as "the Saudi Arabia of natural gas" and by the Toronto *Globe and Mail* as a "prolific monster" with the potential to "rearrange the continent's energy flow."

In the sense of "abnormal to the point of inspiring horror," "monster" is not an inappropriate term here. With every well drilled—and 32,000 wells per year are planned—a couple million gallons of freshwater are transformed into toxic fracking fluid. Some of that fluid will remain underground. Some will come flying back out of the hole, bringing with it other monsters: benzene, brine, radioactivity, and heavy metals that, for the past 400 million years, had been safely locked up a mile below us, estranged from the surface world of living creatures. No one knows what to do with this lethal flowback—a million or more gallons of it for every wellhead. Too caustic for reuse as is, it sloshes around in open pits and sometimes is hauled away in fleets of trucks to be forced under pressure down a disposal well. And it is sometimes clandestinely dumped.

By 2012, 100 billion gallons per year of freshwater will be turned into toxic fracking fluid. The technology to transform it back to drinkable water does not exist. And, even if it did, where would we put all the noxious, radioactive substances we capture from it?

Here, then, are the environmental precepts violated by hydrofracking: 1) Environmental degradation of the commons should be factored into the price structure of the product (full-cost accounting), whose true carbon footprint—inclusive of all those diesel truck trips, blowouts, and methane leaks—requires calculation (life-cycle analysis). 2) Benefit of the doubt goes to public health, not the things that threaten it, especially in situations where catastrophic harm—aquifer contamination with carcinogens—is unremediable (the Precautionary Principle). 3) There is no away.

This year I've attended scientific conferences and community forums on fracking. I've heard a PhD geologist worry about the thousands of unmapped, abandoned wells scattered across New York from long-ago drilling operations. (What if pressurized fracking fluid, to be entombed in the shale beneath our aquifers, found an old borehole? Could it come squirting back up to the surface? Could it rise as vapor through hairline cracks?) I've heard a hazardous materials specialist describe to a crowd of people living in fracked communities how many parts per million of benzene will raise risks for leukemia and sperm abnormalities linked to birth deformities. I've heard a woman who lives by a fracking operation in Pennsylvania—whose pond bubbles with methane and whose kids have nosebleeds at night—ask how she could keep her children safe. She was asking me. And I had no answer. Thirty-seven percent of the land in the township where I live with my own kids is already leased to the frackers. There is no away.

RIVER KILLERS

The False Solution of Megadams

JUAN PABLO ORREGO

DESTROYING WILD RIVERS *to generate electricity is a false solution to humanity's need for energy, with extremely high costs to individuals, communities, and ecosystems. The negative impacts of large dams are multidimensional—degrading watersheds, riparian zones, coastal ecosystems, and even oceans. We must reject the lure of megadams and the energy gluttony they perpetuate, which deters society from deploying distributed power generation at a much smaller scale, consuming less energy, and letting nature produce more beauty.*

❖

Again and again in recent history, humans have rushed headlong to adopt the latest discovery, invention, or technology, embracing them as *the* panacea to solve humanity's problems in a particular field. We have done so with remarkable and mounting shortsightedness, even eagerness, due to the anticipated (and often real) economic windfall associated with the utilization of new technologies. In many cases, the secondary, synergistic, cumulative impacts of the latest miraculous techno-fix have been devastating.

Petroleum-fueled internal combustion engines, nuclear fission, DDT, chlorofluorocarbons, asbestos,... the list is long of "technological wonders" that spawned unintended consequences. The negative effects of the reckless use of these technologies are directly related to the lack of wisdom applied when taking the decision to deploy and use them. What is needed is a systemic and holistic approach to the temporal and spatial/ecological dimensions of technological developments.

Destroying wild rivers with large dams in order to generate electricity is one of the clearer examples of a false solution to humanity's "need" for energy. Modernity has unnecessarily inflated this need; given the severe negative environmental impacts of electricity generation in general, it is amazing how superfluously and frivolously this form of energy is utilized. At this point in human history, our capacity to have blind spots regarding truly life-or-death issues has become one of our most prominent traits.

Large hydroelectric dams—with a height from foundation to crest exceeding 15 meters (49 feet)—are a new technology tied to the development of modern metallic cements, a history that dates back only some eighty years to the building of the Hoover Dam on the Colorado River in 1931. Since then, the number of large dams around the world has increased to more than 45,000, with the largest reaching 300 meters (nearly 1,000 feet) in height.[1] Dams, interbasin transfers, and water withdrawals for irrigation have fragmented 60 percent of the world's rivers.[2] In geological and even human timescales these eight decades represent a very short time span, particularly when attempting to elucidate how this controversial megatechnology stands in terms of its cost-benefit equation.

With the building of the first hydroelectric megadams, the technology was touted as a clean, abundant, renewable, and cheap source of energy. Mounting evidence demonstrates that most of these assertions are ideological. Such claims are biased by the extremely profitable business that surrounds megadams all along their life cycle, which includes design, financing, environmental evaluation, and actual construction with its mobilization of workers, materials, and machinery.

Megadams are river killers. The transformation of watersheds and fluvial ecosystems—crucial components of the planet's circulatory system—into hydraulic artifacts has proven to be extremely costly. Large dams degrade, homogenize, and impoverish rivers' natural dynamics on local to continental to global scales, with serious global biodiversity implications.[3] Most of the vital biological and physiochemical variables of rivers are altered by dams—water and sediment flows, temperature, oxygen content—with resulting upstream and downstream degradation of the food chain. The periodic manipulation of a river due to the operation of a hydroelectric plant results in flow fluctuations from zero water discharge to very large surges. This can, among other deleterious effects, eliminate a river's natural seasonal cycles, even altering irreversibly the physical structure of river channels.

Developers have been particularly blind regarding the vital importance of riparian (riverside) ecosystems and wetlands for the health of rivers and hydrological basins. Rivers are an integrated whole from source to mouth. In fact, rivers and the watersheds that nurture them need to be understood multidimensionally, not linearly as a belt of water from which engineers attempt to reap as much kinetic energy as possible.

The negative impacts of large dams are also multidimensional, degrading in complex ways a watershed's web of life. The capture of sediments within the reservoir via gravity is one example of an uncounted negative consequence. Sediments, both organic—called nutrients—and inorganic, are vital for the food chain. Nutrients, as the name indicates, nurture life, but inorganic elements are also indispensable for certain microorganisms such as diatoms, which are at the base of the marine web of life. Diatoms need the silica provided by rivers to build their exoskeletons.

In my country, Chilean marine biologists have defined Patagonia's coastal ecosystems as "estuarine," a term one associates with rivers rather than oceans. The scientists have concluded that rivers are the main ecological pillar of these coastal ecosystems; they supply both types of sediments, resulting in productivity typical of littoral waters, many times higher than in the interior seas. For this reason, marine fish come to the coast to spawn. Thus, degrading a fluvial ecosystem means degrading coastal ecosystems and even oceans. Killing a river can starve whales hundreds of kilometers away.

Another recent insight is that megadams are important emitters of greenhouse gases, particularly in tropical latitudes. Research by scientists at Brazil's National Institute for Space Research (INPE) has demonstrated that the world's large dams emit annually 104 million metric tons of methane, and approximately 15 percent of total anthropogenic greenhouse gas emissions. This is logical. Due to the putrefaction of organic matter, all bodies of water, natural and artificial, emit methane, which is at least 25 times more powerful than carbon dioxide in its greenhouse effect. Dams submerge soils and drown all forms of life previously existing within the flooded area. The emission effect lasts throughout a dam's life cycle, given that the dammed river keeps bringing organic matter into the reservoir. It is compounded by the fact that by massively destroying photosynthetic organisms, both plants and phytoplankton, dams eliminate carbon sequestration capacity. It is a double punch to the climate. With more than 45,000 large dams around the globe, their overall effect as climate changers has become a planetary problem.

Megadams are also weather changers at the level of basins. The stored water absorbs heat during the day and releases it at night, altering temperature regimes and wind patterns. Reservoirs also become massive evaporative devices, significantly augmenting humidity rates with unforeseeable consequences for flora and fauna. The weight of enormous volumes of water suddenly appearing in a valley can also induce earth tremors, a phenomenon called "reservoir-induced seismicity." These and other documented impacts of megadams are clear, and they undermine the arguments of proponents who tout large-scale hydropower as a renewable, clean, cheap source of electricity.

The negative social consequences of large dams are proportional to their multiple environmental impacts, underscoring an obvious point: The social and the ecological are totally intertwined. One of the most striking discoveries one will find when researching large dams is the uncertainty regarding their collateral damage to human communities. The relevant literature estimates that between 40 and 80 million people have been displaced—in many cases forcibly relocated—due to the building of large dams.[4] China and India, the countries with the most people displaced by megadam projects, are not forthcoming with official information on the subject. Another trick utilized to hide the social impacts of dams is to artificially reduce, on paper, their area of influence. This is the opposite of a systemic analysis. In many cases, Environmental Impact Assessments are contracted, paid for, and edited by dam proponents, and the area considered is a limited footprint around the reservoir. Synergistic and cumulative upstream and downstream effects, and the overall integrity of the hydrological basin, are ignored.

A few years ago, I was among a group of megadam fighters from around the world who met in a school in a small Guatemalan village. For three days we were "stared at" by large portraits hanging from the walls around us of the 378 Maya Ach' children, women, and men who had been murdered by the army to pave the way for the building of the Chixoy dam, funded by the World Bank, the Inter-American Development Bank, and the Italian government.

Since the building of two large hydroelectric dams on the Biobío River in south-central Chile in the late 1990s, the sacrificed region has become the poorest in Chile, with the country's highest suicide rate (triple the national average). In the Upper Biobío, under the ominous shadow of two dams, 113 and 155 meters high, citizens must pay electricity bills that are among the most expensive in Chile, while in Santiago we pay the cheapest, an incentive for the sprawling city to keep growing.

Despite the documented ecological and social effects of mega-hydro projects,[5] dam proponents continue to find receptive ears among government officials. The Three Gorges Dam in China is the most prominent recent example; the monstrous, 660-kilometer-long impoundment dammed the Yangtze River, flooding 258,225 acres, displacing 1.3 million people, and affecting another 10 million. Activists and local people are currently resisting similarly damaging projects such as the proposed Belo Monte dam in Brazil's Amazon and the HidroAysén development scheme in Chilean Patagonia, one of the Earth's last great wilderness regions.

The opposition to this destructive proposal is fierce. In 2005, a Chilean coalition called the Patagonia Defense Council launched the Patagonia Without Dams campaign. Today, thousands of conservationists and 69 organizations from multiple countries are working to stop both the damming of the Baker and Pascua rivers in the heart of Patagonia and the building of an associated transmission line (at roughly 1,200 miles long, largest of its kind in the world).

The Patagonia dams project is promoted by the Italian energy giant Enel, controlled by the Italian government, which retains 32 percent ownership. The antidams campaign has already delayed the project through organizing and litigation. The intense public education efforts of the coalition have been very successful in putting the cultural and ecological value of Patagonia before the eyes of the national and international public. Nearly 80 percent of the Chilean people oppose the project, and many Italian organizations under the "Patagonia Senza Dighe" campaign are confronting Enel at home for its intentions to dam Patagonia's wild rivers.

An unprecedented Chilean citizen's alliance is currently highlighting the megadams' absurdity in terms of energy policy. A thorough technical proposal for a new electrical grid and policy has been drafted; the study demonstrates that the HidroAysén project is not only unnecessary but the worst step the nation could take regarding energy development. Chile has no power deficit at the moment and is blessed with a geography that makes it uniquely positioned to harness renewable sources of energy including solar, wind, geothermal, oceanic, biomass, and small-scale, run-of-the-river hydropower.

It is true that the world's large dams have generated vast amounts of electricity. By 2008, hydropower represented 16 percent of the world's electricity production, with plants in more than 150 countries.[6] It represents over 90 percent of the total electricity supply in 24 countries and more than 50 percent in 63 countries.[7] Canada, the United States, Brazil, China, and Russia account for more than half the world's hydropower generation. Between 1973 and 1996 hydropower generation in non-OECD countries grew from 29 to 50 percent of world production, with Latin America increasing its share by the greatest amount in that period.[8] The trend suggests that good hydropower sites in northern locations like Europe and North America have been largely exhausted, and/or projects are harder to implement due to externalities. Thus the trend to export the megadam business to the South.

In the North, some dams are even being decommissioned, a process that may be more expensive than the original construction. In some cases, the behemoths, with massive amounts of wet sediments behind huge walls of eventually collapsible cement, have become environmental "passives," threats comparable to nuclear dumps. Who is responsible for demolishing river-killing dams, restoring the basin, bringing the river back to life, and helping disintegrated communities heal? Until now this crucial concern has not been evaluated in Environmental Impact Assessments and has not been considered in cost-benefit equations. The hard question after acknowledging the large electrical contribution of megadams is: Has this "cheap" electricity fueled sustainable development with clean energy, or has it powered *overdevelopment* with a source of energy that maims the biosphere, humanity included?

Considering our actual planetary situation, social and ecological, wouldn't it have been much better to have abstained from the use of this energy, blindly seen as clean and cheap, and to have conserved the numerous hydrological basins and wild rivers now harmed by megadams? Wouldn't it have been wise to protect the waters, the biodiversity, the hydrological and atmospheric cycles, the climate, the life of the littoral ecosystems and the seas, the fisheries, the livelihoods, the communities, the local economies, the beauty and grace?

We urgently have to think about energy in a radically different way. We have to assume that humanity's fundamental challenge is not how to *generate more* but how to *curtail demand* and consumption. Then we have to reorient our societies toward the honest quest for the common good and environmental sustainability instead of toward rapacious profits generated from necessary social services. The overwhelming scale of ecological destruction that accompanies large dams is the direct consequence of current patterns of economic growth, of particular modes of so-called development and the concomitant technologies that support it and flow from it.

Large dams are a manifestation, a symptom of a pathological pattern compounded by ignorance and greed. Degradation of watersheds, and of local communities and local economies, is part of the nature of large hydropower dams. Far from being neutral technologies, they orient social processes toward more centralization and concentration of power over natural resources and ecosystems—euphemisms for vital organs of the biosphere—and capital, leading to more authoritarian and inequitable political systems.

We need to totally reject the lure of megadams and the energy gluttony they perpetuate, which deters us from deploying distributed power generation at a much smaller scale, administered at the local level, using all the genuinely renewable sources of energy. This is the path toward humans consuming less energy and less stuff, and toward letting nature produce more beauty and wildness.

While it can be daunting work, defending wild rivers from large hydropower projects is an honor. It is a privilege to become the human voice for all the life of a watershed, and to join a network of people around the world working for similar causes. We dream that soon more and more anti-megadam campaigns will be won, and the lost ones will offer lessons to learn from, making us appreciate the fearsome beauty of the present even more.

BIOENERGY

A Disaster for Biodiversity, Health, and Human Rights

RACHEL SMOLKER

A NEW, GLOBAL *rush to embrace biofuels—for transport, heat, and electricity—is a growing threat to ecosystems, wildlife, human health, and the climate. The trend poses the danger of increased commodification of forests, greater competition between food and energy markets, and even more pressure on the world's rural poor that depend upon local biomass for their energy needs.*

❖

Since humans first learned to manipulate fire, people have used local biomass—including wood, other plant matter, and dried animal dung—for heat and for cooking. Billions of people continue to do so. But now, in addition to these traditional uses there is an unprecedented push for large-scale industrial/commercial bioenergy. This new trend includes refining plant materials (corn, wheat and other grains, sugarcane, soy and palm oil) to make liquid biofuels for transportation and burning plant materials (wood, agricultural residues, municipal waste, etc.) for heat and electricity. Less widely known is the development of plant-based petroleum substitutes for use in bioplastics, biochemicals, inks, fabrics, pharmaceuticals, and other products. Proponents refer to a new "bioeconomy" featuring massive biorefineries that take in millions of tons of plant biomass and convert them into all manner of energy and materials.

But two important questions are often overlooked in the rush toward bioenergy: Where will all that plant biomass come from, and what will the consequences be on ecosystems, wildlife, agriculture, human rights, climate, water, and soil?

TRANSPORT FUELS

Biofuels for transportation—ethanol and biodiesel—have been enthusiastically embraced as "green" alternatives to petroleum fuels, with claims that they would reduce greenhouse gas emissions while reviving domestic fuel production. Brazil is the model country, having already made considerable progress toward replacing petroleum fuels with sugarcane ethanol nationwide. In the United States, ethanol from corn is supported with generous subsidies. Mandated targets for biofuel use have been signed into law in the United States, Europe, and elsewhere.

The negative impacts of this rush to biofuels are already apparent. Brazil's sugarcane ethanol industry is converting vast parts of the delicate Cerrado savanna ecoregion into industrial sugar monocultures—cleared, plowed, sprayed with chemicals, and repeatedly burned over. The appalling work conditions of "sugar slaves" have also been documented. In the United States, expanding corn production for ethanol has resulted in the increased use of synthetic fertilizers (visible in the expanding dead zone in the Gulf of Mexico), former conservation lands being planted with corn, and depletion of freshwater aquifers. Increased demand for corn has also shifted U.S. production away from soybeans, causing production in Brazil and elsewhere to expand to fill the void, often at the expense of tropical rainforests. This sort of "indirect land use change" has been a topic of heated debate, and industry has fought to exclude it from consideration—because when indirect land use is taken into consideration, virtually all biofuels result in more greenhouse gas emissions than does petroleum.

The "poster child" for negative impacts from biofuel production is palm oil biodiesel. Given the recent rate of deforestation and widespread investment in oil palm plantations and biodiesel, it has been estimated that 98 percent of the forest on Borneo and Sumatra —among the most diverse on Earth—could be cut down and replaced by oil palm monocultures by 2022.[1] Emissions from the conversion of Asia's lowland peat forest into palm oil plantations are astronomical, accounting for nearly 8 percent of the global total. The expansion of palm oil is at the root of many human rights abuses in Colombia, Ecuador, and other Latin American countries, as *campesinos* are violently expelled from their lands or even murdered to make way for industrial plantations. Ironically, palm oil plantations are, by U.N. definitions, considered "forests" and increasingly rewarded with carbon finance intended for forest protection.[2]

Public opinion toward biofuels has soured over the years. This shifting attitude was aided especially by a leaked World Bank memo in 2008 that noted how the diversion of food crops into ethanol was a major factor in driving up food prices. With more than a billion people living with chronic malnourishment, U.N. Special Rapporteur on the Right to Food Jean Zeigler in 2007 called the conversion of corn, soybean, cassava, wheat, vegetable oils, and other food into fuel for automobiles, a "crime against humanity."[3] In 2010, the United States nonetheless put nearly a third of its corn crop into ethanol production.

BIOMASS FOR HEAT AND ELECTRICITY

Around the globe, initiatives to reduce greenhouse gas emissions have frequently morphed into policies supporting biomass combustion as it is viewed as the least costly and most adaptable alternative. Electric utilities find it easier to convert large coal-burning facilities to burn a mix of coal and wood chips than to shift to wind or solar power production. Wood is readily available in many locations year-round, and can provide reliable baseload power that can be distributed on the current grid.

Europe, having accepted emissions reduction targets under the Kyoto Protocol, already provides about two-thirds of its so-called renewable energy from biomass—indeed, it accounts for 80 percent of the growth in renewables between 1990 and 2005. The European Commission estimates that 14 percent of the European Union's total energy will be generated from burning biomass by 2020. In the United States, state renewable portfolio standards and regional agreements to reduce emissions similarly favor biomass combustion, and a slew of federal subsidies intended to develop renewable energy are largely directed to biomass burning. Hundreds of new biomass-burning facilities are proposed or under construction around the country, in addition to plans for converting both large coal-burning facilities—and also many small-scale facilities such as schools and hospitals—to biomass energy.

What will they burn? Definitions of "biomass" vary geographically and across different policies, but it can include wood and other plant materials, animal manures, slaughterhouse remains, sewage sludge, municipal solid waste, construction debris, even tires and plastics—virtually anything of remotely biological origin. The bulk of what is burned as biomass, however, is wood, followed by vegetable oils.

Biomass generally has a low "energy density," far less than conventional fossil fuels, and thus it takes far more wood, for example, than coal, to generate equivalent energy. On average (depending on facility efficiency and wood moisture levels), it takes approximately 1.5 tons of wood to generate just a megawatt-hour of electrical power. The demand for wood is therefore huge, even for a moderately sized 50-megawatt facility, and will be ongoing for the lifetime of the facility. Biomass plants generally operate at only 25–30 percent efficiency, meaning that for every four trees burned, the energy content of only one is actually retrieved. Smoke from all four burned trees (or animal remains, or municipal waste) is emitted into the air, however, posing a public health threat. Nitrous oxides, sulphur dioxides, and volatile organic compounds are emitted, along with large quantities of particulates. The U.S. Environmental Protection Agency regulations restrict but do not wholly prevent emissions of all particulates, particularly "fine particles," of 2.5 microns and smaller. Emerging science indicates these are extremely dangerous because they can lodge deeply in the lungs, enter the bloodstream, and result in numerous negative health effects, especially cardiopulmonary disease and cancer.

When communities oppose biomass facilities it is usually due to concerns about plant emissions, but the impact on forests may ultimately pose the greater risk to human health. Forests are essential to maintaining life on Earth, and as we face the consequences of climate change, they are one of our best defenses—sequestering carbon, harboring biodiversity, and helping regulate hydrological cycles. Human activities have already resulted in a massive loss of forest cover in modern times, but instead of restoring and protecting forests we are rapidly escalating the pace of destruction, now in the name of producing "renewable energy."

In the state of Massachussetts, activists battled proposed biomass facilities that would have burned, in total, at least 2.4 million tons of wood per year, hauled in on 600 logging truck trips per day, to increase the state's generation capacity a mere 1.2 percent. Like other thermoelectric facilities, these facilities would require millions of gallons of cooling water, much of which would be lost as vapor, and the rest (heated and contaminated) dumped back into the waterways. The state is revising its regulations in response to fierce opposition.

Ohio recently considered (and at least partly rejected) a whopping 2,400 megawatts of biomass-generated electricity capacity, mostly through cofiring with coal in nine different power plants. Just one of the facilities, First Energy's Burger plant, would have burned more than 3 million tons of wood per year, double the current 1.7 million tons produced anually by the state's timber industry. Based on data and projections from the Energy Information Agency, if the United States were to adopt a Renewable Energy Standard mandating 25 percent renewable energy by 2025, as has been proposed, biomass energy production would expand to require the equivalent of clear-cutting 50 million acres of forest by 2030.

The premise that burning biomass is clean, green, and carbon neutral (the carbon neutral myth enables developers to cash in on subsidies and credits) is simply false. Burning biomass releases more carbon dioxide per unit of electricity generated than does burning coal or natural gas. Additional emissions result from logging operations, soil disturbance, and biomass transportation. Trees may eventually regrow and resequester carbon, but only after many years or even decades—a "carbon debt" time frame that is hardly meaningful if we are to address global warming. Most scientists believe we need to reduce greenhouse gas emissions immediately, not increase emissions above and beyond even what would occur if we continued burning coal, in hopes that trees of the future will reabsorb that extra carbon dioxide. Moreover, many forests are already diminished from previous overcutting, and they are now declining due to the impacts of climate change.

WOOD CHIPS AS GLOBAL ENERGY COMMODITY

A new and rapidly expanding international trade in wood chips and pellets has sprung up, in large part to satisfy growing European demand. As an example, energy company RWE's Tilbury facility in the United Kingdom was recently approved to convert from coal to biomass. This facility alone will burn more than 7 million tons of (mostly imported) wood pellets per year. In combination with other existing and proposed biomass facilities, demand within the U.K. would rise to over 60 million tons per year, more than six times annual U.K. production of wood.

Europe's forests are already severely over-exploited, so most wood chips and pellets are imported from around the world. For example, MagForest, a Canadian company operating in the Republic of Congo, will soon ship 500,000 tons of wood chips annually to Europe. IBIC Ghana Ltd. claims it can ship 100,000 tons of tropical hardwood and softwood a month from Ghana for bioenergy. Sky Trading, a U.S. company, is offering to supply up to 600,000 tons of wood chips for biomass from the United States or Brazil. Green Energy Resources, based in Nevada, has stated a goal of "supplying 20 percent of the European demand for wood chips by 2015." Brazil's International CMO Business Biomass says it is dedicated to reducing coal use and can obtain wood chips from Brazil, Chile, Uruguay, and Argentina to supply the European energy market.

This huge new demand for woody biomass is also spurring the creation of more industrial tree monocultures. For example, a South Korean company has applied for a 200,000-hectare concession in Indonesia to produce wood pellets for "green energy." The U.K. firm Carbon Positive has entered a joint venture to develop 160,000 hectares of tree plantations for bioenergy in Indonesia, including in West Papua. Conservation International is helping the Indonesian company Medco to develop plantations for wood pellets—up to 300,000 hectares, mainly in West Papua. In the United States, Arborgen is developing genetically engineered tree varieties claiming they will provide "more wood on less land." As the chief executive officer of a German energy company has said: "Wood is very quickly becoming an important part of the energy mix and in a few years will be a global commodity much like oil."

The threat to forests is dire. In the United States, industry groups are calling for more access to public and private lands for biomass extraction. They claim that harvesting biomass will "protect" forests by ensuring they remain profitable to landowners. Others call for access to beetle-damaged forests, claiming that it is better to "make good use" of the dead and dying trees rather than allowing them to decay; but harvesting from these damaged forests can worsen their condition and risks spreading the beetles. Developers of new biomass-burning plants often claim they will use "only wastes and residues" (i.e., from past logging operations), even though competition for these materials is already great. Maps of the sourcing areas for many proposed and existing biomass burners show them overlapping with one another, indicating there will be fierce competition for scarce wood supplies, far beyond what "wastes and residues" can supply. Some biomass boosters advocate for removing more woody material from logging sites, taking tops and limbs that would "otherwise decompose." But removing so much material is akin to mining forest soils; without large quantities of organic material left to decompose on the forest floor after harvest, soils are stripped of nutrients and left exposed to erosion. Finally, the claim is often made that many forests in the United States need to be thinned in response to insect damage, to prevent burning and protect homes from wildfire damage. But studies have indicated that forest thinning results in more light penetration, drying out soils and vegetation and making forests more—not less—vulnerable to fires.

All of these are thinly veiled excuses that have little to do with protecting forests or public health, or even providing "green jobs." At base they are about getting access to the massive amounts of wood needed for this lucrative, heavily subsidized forest incineration industry.

THE NEW BIOECONOMY

In addition to biofuels for transportation and biomass combustion for electricity and heat, proposals abound for developing plant-based alternatives to petroleum for a host of applications, including products such as plastics and chemicals. One such example is a recent partnership between Dow Chemical and Japan's Mitsui to construct a large facility in Brazil that will use sugarcane ethanol to produce bioplastics for packaging. Along similar lines, Dow has partnered with Brazil's Crystalsev to develop a large facility to produce biobased polyethelene. All of these initiatives require large amounts of plant material, water, and land.

In the end, the push to replace fossil fuel energy with plant biomass is fueling the fires of injustice and inequality. Billions of people around the world rely entirely on firewood for cooking, and on animals for energy and transportation. The vast majority of residents in these traditional, biomass-dependent communities do not own cars capable of burning ethanol, or have electricity in their homes, yet they are now under siege as the new bioeconomy creates new demands for land, water, soils, and biomass. Assessments of "global biomass availability" portray vast areas—largely in the Global South, where growing conditions are favorable—as "marginal" and available for growing and extracting biomass. In reality, peasant farmers, pastoralists, and others on the margins of the global economy are dependent on those lands. And they are already facing eviction and violence in the face of increasing pressure for access to their lands—in the name of renewable energy.

OIL SHALE DEVELOPMENT

Looming Threat to Western Wildlands

GEORGE WUERTHNER

OIL SHALES, IF *they live up to proponents' expectations and can be produced commercially, could change the economic and political fortunes of the United States and transform the geopolitical map of the world. But any large-scale effort to exploit oil shales will threaten wildlife habitat and water quality, and exacerbate climate change.*

❖

O il shales are rock–like, hydrocarbon–bearing formations. The term is somewhat of a misnomer because oil shale does not contain any crude oil, and is not necessarily associated with actual shale rocks either. (Oil shale should not be confused with shale oil, which is crude oil found in shale formations such as the Bakken Formation in North Dakota.) Like Alberta's tar sands, oil shales, when heated and processed through a chemical process known as "pyrolysis," can produce an oil–like substance. The liquid resulting from oil shale is not a direct substitute for conventional oil, but it can be used to produce diesel, jet fuel, and kerosene.

Oil shales were created millions of years ago as organic matter accumulated in a mix of sand and mud on the bottom of inland seas. The resulting substance, known as "kerogen," is similar to oil but without having been subjected to the same heat and pressures. In a sense, oil shale is oil in waiting. If we could add many millions of years of pressure and heat, the kerogen would be converted into conventional oil.

The extraction and processing of oil shale is essentially an attempt to speed up the formation of oil by investing energy to assist the conversion— but this raises the cost of any oil produced. Although geologists and oil companies have known about oil shales for decades, the costs of extracting and processing them made them uncompetitive with conventional oil.

Oil shales are found in locations around the globe including Scotland, Germany, Estonia, Russia, China, and Brazil. Oil shale deposits also contain natural gas and natural gas liquids. Since oil shale can be burned without any additional processing, it is sometimes used like coal as a fuel for power generation. Estonia, for instance, uses oil shale for the majority of its electrical generation.

The largest known deposits of oil shale are found in the Green River Formation, which sprawls across northwest Colorado, southwest Wyoming, and northeastern Utah. Some of the richest parts of the formation are found in northwest Colorado at a depth of between 1,000 and 2,000 feet. These deposits contain enough hydrocarbons to theoretically produce about a million barrels of oil equivalent per acre. In comparison, the best tar sands formations in Alberta produce around 100,000 barrels of oil equivalent per acre.[1] The size of "in-place" global oil shale deposits is not well known, but it's safe to say there are trillions of barrels of oil equivalent.[2] ("In-place resources" are the total amount of oil thought to be contained in the formation regardless of technical or economic recoverability. What can be extracted is always considerably less due to technical, economic, political, and other limitations.) The most recent U.S. Geological Survey (USGS) estimates suggest there may be over 4 trillion barrels of in-place oil resources in the Green River Formation. USGS studies found that the Piceance Basin of northwest Colorado alone had an estimated 1.5 trillion barrels of in-place oil resources. The USGS also estimated in-place resources of 1.3 trillion barrels of oil in the Uinta Basin of Utah and 1.4 trillion barrels of oil in the Green River Basin of Wyoming.[3] The Energy Information Administration estimates that about 800 billion barrels of oil are potentially recoverable from U.S. oil shales with current technologies.[4]

Since the United States currently consumes roughly 7 billion barrels of oil annually, some people view oil shale development as a potential panacea for meeting the nation's future energy needs. The military, in particular, sees oil shales as a potential contributor to energy independence and national security.[5] So far, however, economic viability has been elusive. In the 1970s there was a short-term rush to develop oil shale commercially. Rifle, Parachute, Rangely, and Meeker in Colorado briefly became boomtowns. But in 1982 Exxon shut down its oil shale operations, and the boom came to an end. More recently, new technologies—many pioneered in the Alberta tar sands operations—combined with higher energy prices have generated renewed interest in oil shales.

That original oil shale boom produced excessive environmental impacts. Oil shale was mined like coal, the rock was crunched and the kerogen removed; then, through a water-intensive process known as hydrogenation, the resulting substance was refined into fuels. The main drawback of this process was that it produced mine tailings from which toxic materials could leach into ground and surface waters, or blow into surrounding countryside. The costs (including the mitigation of environmental hazards) made oil shales too expensive to develop.

The recent rise in the price of oil has brought new attention to oil shales, and new technologies may be able to reduce the environmental impacts. One of these innovations is In-situ Conversion Process (ICP), in which the kerogen is heated in place below ground, thereby reducing the need to move rock to the surface and eliminating the piles of tailings left over after the mining and refining process. The heated kerogen is liquefied and then pumped to the surface, where it undergoes processing into various fuels.

Shell Oil Company has refined ICP further: Electric heaters are placed in the oil shale formations for two to three years, which gradually melts the kerogen so it can be pumped to the surface. At the same time a freeze wall is created as a barrier to prevent kerogen from migrating outside the extraction zone and potentially polluting surrounding aquifers. All this takes a lot of energy, and whether it is viable at commercial scales remains to be seen. Shell reports that it can get an energy return of three to four barrels of oil for the investment of one barrel of oil's energy, a low net energy ratio compared to conventional oil production.

There are also huge up-front costs in developing such unconventional hydrocarbons, with oil price volatility being the biggest obstacle to commercial development of oil shales. Since there is a long lead time between when the hydrocarbon-bearing formations are first exploited and the final production of fuels, a sudden change in oil prices could make or break commercial operations.

Given the world's demand for oil and the decline of known conventional sources of oil, it's not difficult to imagine that there will be a price point where oil shale may be economically viable to recover at commercial scale. Much of the surface lands over the Green River oil shale formation are owned by the federal government and controlled by the Bureau of Land Management. If a new oil shale boom occurs, there will be a rush to lease these public lands, with adverse impacts on fragile desert ecosystems: new roads, expanded development, water pollution, dust pollution, and a general degradation of one of the most scenic, iconic, and wildest landscapes in the West.

Despite the negative effects on wildlands and wildlife of this region, the biggest global threat from oil shale–derived fuel results when we burn those fossil fuels in our vehicles. Like tar sands, the abundance of oil shales—should they ever prove profitable to exploit—could potentially slow global efforts to rein in greenhouse gas emissions. An oil shale boom would open Pandora's Box. It is a box best kept closed.

GAS HYDRATES

A Dangerously Large Source of Unconventional Hydrocarbons

GEORGE WUERTHNER

GAS HYDRATES ARE *a frozen form of methane found in Arctic regions as well as under the seabed. They are one of the most abundant hydrocarbons on the planet and may contain twice the energy of all other hydrocarbons combined. Gas hydrates are seen as a potentially new energy source in countries like Japan, where other sources of fossil fuel are not abundant. Commercially exploiting gas hydrates on any significant scale may prove to be extremely challenging—but if successful, it would prolong our dependency on fossil fuels and contribute to ever-growing greenhouse gas emissions.*

❖

Under scrutiny as one of the newest energy sources that could help satisfy global energy demand, gas hydrates represent an immense source of methane (the main component of what we call "natural gas"). Gas hydrates are frozen, water-based crystalline solids that trap methane inside; they form at high pressures and low temperatures. Although these deposits look like ice, they turn to water and gas when pressure is relieved or temperatures increase. Massive amounts of gas hydrates exist in deep-sea sediments, on land associated with Arctic permafrost, and sometimes in deep-lake sediments, such as under Lake Baikal in Russia. The majority of ocean-floor gas hydrates are found at depths of more than 1,500 feet (500 meters).

Under hydrate conditions, gas is extremely concentrated. One unit volume of methane hydrate at a pressure of one atmosphere produces about 160 unit volumes of methane gas—thus gas hydrates are very energy-dense reservoirs of fossil fuel. The quantity of methane in gas hydrates worldwide is poorly known, but has been estimated by the U.S. Geological Service (USGS) to be equal to twice the amount of carbon held in all other fossil fuels—all the oil, gas, and coal combined—on

Earth.[1] While highly speculative for a fossil energy resource that has essentially zero commercial production at present, interest in hydrates has increased in some parts of the world where other sources of energy are less available or more expensive.

The U.S. Geological Survey estimates that the United States alone holds potentially 200,000 trillion cubic feet (Tcf) of natural gas in gas hydrate deposits.[2] To put that in perspective, in 2010 the United States consumed around 24 trillion cubic feet of natural gas. However, only a small proportion of global hydrate resources may ever be developed due to the technological challenges involving temperature, pressure, environmental protection, and other factors—all of which add to the costs (and energy) required to produce the gas.

The largest known deposit of gas hydrates lies on the continental shelf of the United States between New Jersey and Georgia. The Blake Ridge gas hydrate deposit occurs off the coasts of North and South Carolina, where the USGS estimates there may be 1,300 trillion cubic feet of methane gas.[3] Another promising location for U.S. gas hydrate development lies in the Gulf of Mexico, where the Department of the Interior has estimated the region contains 21,000 trillion cubic feet of methane.[4]

In the near term, the most accessible hydrate deposits occur in the Arctic. Recent drilling on Alaska's North Slope suggests there may be a minimum of 85 trillion cubic feet of undiscovered, "technically recoverable" (i.e., recoverable with current technologies, but without regard to economics) gas resources within gas hydrates in northern Alaska[5]; meanwhile, a USGS estimate puts the possible total in-place gas hydrates for northern Alaska at more than 590 trillion cubic feet.[6] Thus these coastal areas and the Alaska North Slope potentially possess enough gas to meet U.S. needs for decades or centuries—if economical means of extraction can be developed. Other large concentrations occur in the Mackenzie River Delta in Canada's Northwest Territory, and in China, India, Japan, and Siberia, among other areas.

Gas hydrates are stable only within a narrow range of temperature and pressure. Under ideal conditions gas hydrates can form a cemented imperious layer that further traps more gas, creating a significant accumulation zone for methane. There is some evidence that changes in pressure and temperature over gas hydrate sediments can precipitate releases of great quantities of methane. Although controversial, some scientists believe that ancient fluctuating global temperatures may have precipitated numerous huge releases of methane into the atmosphere—leading to global warming that could have possibly contributed to past extinctions.[7] The timing of a massive release of methane is speculated to have been at least one factor in the Permian–Triassic extinction event that caused the greatest mass die-off of species ever recorded. It has been called the "Mother of all Extinctions," with 96 percent of all marine species and 70 percent of terrestrial vertebrate species becoming extinct. Even without causing major extinctions, methane releases are implicated in global climate change. Another global warming event about 55 million years ago is also suspected to be a consequence of the sudden release of massive amounts of methane that had been trapped under the seafloor as gas hydrates.[8]

According to the National Energy Technology Laboratory, the total amount of carbon stored in gas hydrate deposits amounts to many thousands of gigatons, greatly exceeding the quantity of carbon that currently resides in the atmosphere. Such figures give credence to concerns that current global temperature rise may start a chain reaction whereby additional methane, presently frozen beneath the sea and Arctic permafrost or activated from northern wetlands, could be liberated. Since methane is 20 times more effective (over a hundred years) at trapping heat than is carbon dioxide, even a small amount of additional methane could lead to rapid temperature rise, which in turn may trigger even further releases of methane.

Besides the phenomenon of methane release as unintentional geohazard, there is a real interest in gas hydrates as an energy resource. Recent experimental exploration drilling has demonstrated that certain gas hydrates may be exploitable using existing drilling technology and equipment—suggesting potential for commercial viability, albeit at low net energy returns.[9] One promising technique being tested on Alaska's North Slope involves injecting carbon dioxide into hydrate structures, resulting in the swapping of carbon dioxide molecules for methane molecules in the solid-water hydrate lattice, the release of methane gas, and the permanent storage of carbon dioxide in the formation.[10] The gas hydrate deposits that hold the most potential for commercial viability are located in the Gulf of Mexico and in the Alaska North Slope, where existing oil development technology and equipment make them attractive for future exploitation.[11]

New combinations of drilling technology (i.e., hydraulic fracturing with horizontal drilling) have, at least for the short term, precipitated a natural gas production boom in shale formations around North America that has reduced gas prices, likely pushing off the day when offshore gas hydrates are viewed as commercially viable—at least in the United States. In other parts of the world less endowed with fossil fuels, gas hydrates are being more actively explored as a potential source of energy. Given the fact that huge quantities of gas hydrates are possibly available to be tapped, and that methane burns cleaner than coal, oil, and other potential fuels, it is likely that there will be a major push to find economic means of utilizing gas hydrates sooner or later.

Gas hydrates offer an enormous tempting target for future energy production, but it's an open question how much of the gas can ultimately be extracted given the major technological and environmental (and ultimately economic) challenges involved. Moreover, a gas hydrates drilling rush could be dangerous in that the perceived abundance of another hydrocarbon resource may undermine the urgent need to develop renewable energy sources. And if such a drilling boom comes to pass in reality, it may exacerbate climate chaos, degrade marine and terrestrial habitats, and contribute to the delusion that perpetual growth is possible on a finite planet.

REGULATORY ILLUSION

BRIAN L. HOREJSI

THE 2010 DEEPWATER HORIZON *oil spill in the Gulf of Mexico briefly focused attention on how the oil and gas industry exploits public resources with little or no accountability. But the larger problem of how corporations and governments engage in a charade of regulation proceeds largely unnoticed. This sham regulatory process has failed to stem the large and growing ecological, social, economic, and democratic costs thrust on the public.*

❖

There are few people in North America who were not aware of, and to some degree disturbed by, the oil slick that began squeezing the aquatic and terrestrial life out of the Gulf of Mexico as a consequence of the 2010 BP Macondo well (Deepwater Horizon) drilling disaster. As was the case with the 1989 Exxon Valdez spill in the Gulf of Alaska, the full extent of the damage will take decades to manifest itself.

If anything positive can be gleaned from this catastrophe, it is that some of the American people and some of the more progressive media are beginning to realize that the oil and gas industry "owns" American government "regulators," a failure of democratic governance that stretches back at least to 1981 when Ronald Reagan brought his deregulation club to Washington. Other realities are being exposed as well, foremost among them that the oil and gas industry makes an awful lot of money by drastically minimizing the risks of practices that have severely damaged—quite likely for a very long time- nearly continent-size ecosystems.

None of these observations are new to people who monitor or investigate the oil and gas industry. The citizens and organizations that work to expose the truth about the industry are not trying to be secretive, but their message has been muted; for decades now, they have been exposed to a crippling campaign of exclusion by mainstream media. Industry watchers have suffered from media suppression, ridicule, and abuse, as well as political, social, and economic persecution fueled by a constant barrage of industry misrepresentation.

It is a common deception—carefully constructed like a house of smoke and mirrors by "regulators," politicians, and the oil and gas industry—that a fair, systematic, scientifically legitimate, and deliberative process exists through which a company must proceed in order to drill a well or lay pipe, whether it be in the rolling hills of Wyoming or the rippling waters off Louisiana. The public believes that regulators, the men and women who wear the title *public servant*, engage in objective and inclusive analysis of the ecological, economic, and social impacts of any proposed development in order to gauge the merits of an application. The reality, however, is that the oil and gas industry, collaborating with governments swayed by campaign funding, has hijacked what should be a legally protected, scientifically sound, accountable, and highly public process.

Just months after the Deepwater Horizon disaster, a major *Washington Post* article by journalists Juliet Eilperin and Scott Higham described a cozy decades-long relationship between the oil industry and the federal agencies that were supposed to be regulating it.[1] The story begins with the 1970 National Environmental Policy Act (NEPA), which gave government agencies authority to require an environmental impact statement for drilling (and related) activities. The oil industry howled, and the intense resistance led to regulators becoming increasingly willing to grant exemptions. In 1978, under President Carter, the White House's Council on Environmental Quality allowed agencies to make "categorical exclusions" from the law within certain guidelines. In practice, this flexibility led to many projects being substantially exempt from environmental assessment and reporting. As Eilperin and Higham describe:

Under Clinton, categorical exclusions granted in the central and western gulf rose from three in 1997 to 795 in 2000. During the Bush administration, [the Minerals Management Service] granted an average of 650 categorical exclusions a year in the region. The number of categorical exclusions dipped to 220 during the Obama administration's first year. One went to BP's Macondo well.

Short-circuiting environmental impact assessment quickly spread to other agencies, as well as on to land. In fiscal years 2006 and 2008, the

Bureau of Land Management granted at least 6,087 categorical exclusions for applications to drill on land in the western United States. The BLM even hired more staff to accelerate approvals.[?] Imagine if you expected to win—and in fact did win—the lottery nearly every time you bought a ticket. That is the equivalent of the "permitting system" to which the oil and gas industry has become accustomed. A reasonable observer would call that a rigged and corrupt system. On the other hand, you may, like many Americans, believe that ecological viability is important. You may believe that the American people have legitimate expectations and visions for public resources that include fully protected, ecologically functional wild land and water systems. You may argue for conservation options to satisfy that vision, and expect that your participation in democratic advocacy for the common good will be treated fairly as you make your case before those same regulators. Yet, unlike those who seek to profit from exploiting public lands, you will be lucky to win that lottery more often than once or twice every hundred times you play, perhaps only stopping or mitigating a handful of proposed energy projects during the course of a conservation career.

For the public, a victory would be to maintain the functional integrity of roadless landscapes, keep wildlife habitat intact, protect seashore and inland wildlife refuges, safeguard shrimp beds, or guarantee in perpetuity, as best we can, ecosystem viability. It might take the form of denying an application because the environmental impact statement was grossly inadequate. Or it could mean rejecting industry proposals to extract energy resources from beneath a public landscape considered by citizens to be of great social or ecological value. Americans are entitled to be angry that these kinds of victories are rare, that extractive industry is favored over conservation time after time, location after location, year after year.

This is the North American regulatory system we have been hamstrung by. It is virtually identical from state to province to federal jurisdiction; it has "authorized" the drilling of at least 500,000 wells on land and another 50,000 wells off shore. The real world ecological impacts of these actions are severe and growing, touching every part of the continent and affecting the chemistry and temperature of the global atmosphere. The regulatory apparatus necessary to constrain the ecological impacts of the oil and gas industry have failed, largely because the political will to substantively —not just cosmetically- -correct the situation is effectively zero. Sadly, the might of American citizens lies largely dormant.

We readily can envision the oil slick that polluted the Gulf of Mexico as a result of possible negligence by BP and the U.S. Minerals Management Service—an oil slick hundreds of miles long, not unlike an iceberg with a huge flotilla of loosely coalesced oil somewhere below the surface, large enough to engulf entire states. Recall the environmental destruction we know was caused: dying birds coated in brown gunk; fish gasping and flopping; plants being choked by a sickly brown dressing; turtle, frog, fish, and birds' eggs, young, and adults dying from contact with toxic, oxygen-choking oil.

That spill was notable not just for its scale, but because much of it was visible to the human eye, and because, for a brief time at least, the attention of the public and the media were focused on the oil and gas industry. Now think of this: What if the oil slick had been invisible to the untrained eye? What if the Gulf's living diversity just began to collapse, the fish, birds, and marine mammals gone, dramatically reduced in number, or absent from places they'd frequented before? Would anyone have been there to film and document the disaster, to take measurements, to monitor, to record, to speak up?

And yet that is what is happening to the natural world, both on land and in the oceans everywhere we look. The everyday practices of individuals and corporations, including the fossil-energy-extraction industry, quietly kill, choke, manipulate, degrade, and deplete the ecological vitality of ecosystems the way a catastrophic oil spill does, although perhaps more slowly. But virtually no one, particularly those whom we thought were there to protect our interests, has been paying attention.

The assault on the biodiversity of North America by the energy industry is not new; but it has certainly ratcheted up in recent years, particularly at the turn of the twenty-first century. Powerful people such as former vice president Cheney have increasingly wielded political influence over the regulatory process on behalf of the oil and gas industry. These interventions generally are intended to eliminate options that could protect ecosystems, and they lead to two destructive outcomes: essentially unfettered access by the industry to public resources, and erosion and avoidance of regulatory and accountability procedures.

Most North Americans, even those living near the affected lands, have been and remain unaware of the land-based ecological equivalent of an industrial oil "slick" that has spread across vast areas of North America—

from the coalbed methane fields of eastern Montana and Wyoming, to the shale-bed fields of North Dakota and Pennsylvania, to the conventional oil and gas fields of Texas, Utah, Alberta, and Prudhoe Bay, Alaska.

The cumulative impacts of roads, pipelines, pits and berms, leases, compressors, vehicles, power lines, wells, processing plants, refineries, and other associated infrastructure remain mostly out of sight and out of mind to the broad public. This maze of infrastructure is obvious to some people, but the impaired function of a tattered ecosystem is rarely obvious—and people across the continent remain uninformed and unaware as the invisible "slick" grows by the day.

Like the changes necessary to reverse the growth of greenhouse gas emissions, only drastic regulatory intervention—honest and true reform—can begin to reverse the industrial takeover of our democracy and the depletion of the Earth's ecological life-support systems. A regulatory process built on resistance to growth and consumption will be our only salvation.

RETOOLING THE PLANET

The False Promise of Geoengineering

ETC GROUP

THE NOTION OF *using geoengineering—large-scale intervention in the Earth's oceans, soils, and/or atmosphere to correct the unintentional harm human activity has done to the climate—is moving from fringe idea to mainstream discussion. Powerful corporations and governments of developed nations, which are responsible for 90 percent of historic greenhouse gas emissions, are the ones with the budgets and technology to execute geoengineering schemes. There is no reason to trust they will have the rights of more vulnerable states or peoples in mind, and the risk to the planet's ecological integrity is great.*

❖

It is beyond dispute that cumulative, local interventions in ecosystems can bring about planetary-level effects. That's why we have human-induced climate change. However, another notion is quickly gaining ground: that we can use geoengineering to purposefully intervene to correct the unintentional harm human activity has done to the climate.

Geoengineering is the intentional, large-scale intervention in the Earth's oceans, soils, and/or atmosphere, especially with the aim of combating climate change. Geoengineering can refer to a wide range of schemes, including blasting sulfate particles into the stratosphere to reflect the Sun's rays, dumping iron particles into the oceans to nurture carbon-dioxide-absorbing plankton, firing silver iodide into clouds to produce rain, genetically engineering crops so their foliage can better reflect sunlight, and various other strategies to affect solar radiation management, remove and sequester carbon dioxide, or modify weather.

University of Calgary physicist and geoengineering advocate David Keith describes geoengineering as "an expedient solution that uses additional technology to counteract unwanted effects without eliminating their root cause."[1] In other words, geoengineering uses new technologies to try to rectify the problems created by the use of old technologies, a classic techno-fix.

Amidst growing public unease and increasing concentrations of carbon dioxide in the atmosphere, Organization for Economic Cooperation and Development (OECD) countries can either adopt socially responsible policies to cut fossil fuel use and consumption or hope for a "silver bullet" array of techno-fixes that will allow them to maintain the status quo and dodge the consequences. Not surprisingly, the silver bullet option—most clearly embodied in the form of geoengineering—is gaining momentum. Also not surprising is that the states in the global North, which are responsible for almost all historic greenhouse gas (GHG) emissions and have either denied climate change or moved slowly toward substantive policy measures to address it, are the ones warming most quickly to the geoengineering option. And they will have de facto control over its deployment. Only the world's richest countries are likely to muster the hardware and software necessary to attempt rearranging the climate and resetting the thermostat. Once the smog clears, the major private-sector players in geoengineering will probably be the same energy, chemical, forestry, and agribusiness companies that bear a large responsibility for creating our current climate predicament—in effect, the same folks who geoengineered us into this mess in the first place.

Choosing geoengineering flies in the face of precaution. We do not know if geoengineering is going to be inexpensive, as proponents insist—especially if geoengineering doesn't work, forestalls constructive alternatives, or causes adverse effects. We do not know how to recall a planetary-scale technology once it has been released. Techniques that alter the composition of the stratosphere or the chemistry of the oceans are likely to have unintended consequences as well as unequal impacts around the world.[2]

The governments that are quietly contemplating funding geoengineering experimentation are the ones that have failed to pony up even minimal funds for mitigation or adaptation action on climate change. Indeed, in some quarters the MAG approach (Mitigation, Adaptation, and Geoengineering) is already being vigorously promoted.[3] These governments will eagerly divert climate change funding away from climate change mitigation and adaptation toward geoengineering if given the opportunity. After all, they can spend the money on their own scientists and corporations to launch initiatives that are more likely to benefit their part of the world. There is no reason for the peoples of most of Africa, Asia, and Latin America to trust that the governments and industries of the biggest carbon-emitting states will protect their interests. Without public debate and without addressing the inequalities between rich countries and poor countries—in terms of both historical responsibility for climate change and the potential impacts of any techniques deployed to address it—geoengineering could be considered an act of geopiracy.

Geoengineering has long been on the table as a possible response to climate change. As early as 1965, the U.S. President's Science Advisory Committee warned, in a report called "Restoring the Quality of Our Environment," that carbon dioxide emissions were modifying the Earth's heat balance.[4] That report, regarded as the first high-level acknowledgment of climate change, went on to recommend a suite of geoengineering options. As James Fleming, the leading historian of weather modification, wryly noted: The first ever official report on ways to address climate change "failed to mention the most obvious option: reducing fossil fuel use."[5] Forty years after the release of the Science Advisory Committee's report, everybody, including the sitting U.S. president, was talking about global warming. Scientists warned that the temperature rise on the Arctic ice cap and Siberian permafrost could "tip" the planet into an ecological tailspin, and the U.S. Congress agreed to study a bill that would establish a national "Weather Modification Operations and Research Board."

The current debate over engineering the Earth's climate can be traced to a paper[6] coauthored by the late Dr. Edward Teller, the Nobel laureate responsible for the hydrogen bomb and one of the most politically influential U.S. scientists in the latter half of the twentieth century. Teller lent his support to geoengineering when he and two colleagues submitted their paper to the 22nd International Seminar on Planetary Emergencies in Erice, Sicily, in 1997. While the authors did not present their views as being endorsed by the U.S. government, their work was conducted at the Lawrence Livermore National Laboratory, under contract with the U.S. Department of Energy.

In 2002 another Nobel laureate, Paul J. Crutzen, who won his prize for pioneering work on the ozone layer, offered grudging support for geoengineering in the journal *Nature*. Since we're living in the "anthropocene" era when humans are increasingly affecting the climate, Crutzen suggested, our future "may well involve internationally accepted, large-scale geoengineering projects."[7] Also in 2002, Teller (who worked for the U.S. Department of Energy), along with colleagues Roderick Hyde and Lowell Wood, submitted an article to the U.S. National Academy of Engineering in which they argued that geoengineering—not reduction of GHG emissions—"is the path mandated by the pertinent provisions of the UN Framework Convention on Climate Change."[8] In 2005, another high-profile climatologist, Yuri Izrael, former vice-chairman of the Intergovernmental Panel on Climate Change and head of the Moscow-based Institute of Global Climate and Ecology Studies, wrote to Russian president Vladimir Putin outlining a proposal to release 600,000 tons of sulfur aerosol into the atmosphere to take a few degrees off global temperatures.

Paul Crutzen returned to the debate in 2006 when he wrote an editorial in the journal *Climatic Change* calling for active research into the use of sulfate-based aerosols to reflect sunlight into the stratosphere in order to cool the Earth.[9] Crutzen, a professor at the Max Planck Institute for Chemistry in Germany, opined that high-altitude balloons and artillery cannons could be used to blast sulfur dioxide into the stratosphere, simulating a volcanic eruption; the sulfur dioxide would convert to sulfate particles. The cost, he reckoned, would run between $25 and $50 billion per year—a figure, he argued, that was well below the trillion dollars spent annually by the world's governments on defense. Crutzen noted that his cost estimates did not include the human cost of premature deaths from particulate pollution. Such tiny reflective particles could be resident in the air for two years. Crutzen willingly acknowledged that his was a risky proposition and insisted that it should be undertaken only if all else fails. He went on to add that the political will to do anything else seemed to have failed already.

In late 2006 NASA's Ames Research Center convened a meeting of geoengineering advocates to explore options, with Lowell Wood presiding. "Mitigation is not happening and is not going to happen," the physicist reportedly told the group. The time has come, he argued, for "an intelligent elimination of undesired heat from the biosphere by technical ways and means." According to Wood, his engineering approach would provide "instant climatic gratification." From that meeting came the beginnings of a campaign to secure funding for geoengineering techniques.

Current support for geoengineering has come from scientific and political circles, as well as mainstream media. Once a few prominent climate scientists had endorsed geoengineering as a scientifically credible endeavor, publishing in the field exploded both in scholarly journals (almost a fivefold increase) and in the popular press (a 12-fold increase).[10] It is now politically correct to talk about geoengineering as a legitimate response to climate change, a credibility shift that the *New York Times* called a "major reversal."[11] John Holdren, Chief Science Advisor to U.S. President Barack Obama, has conceded that the administration is considering geoengineering options to combat climate change,[12] and U.S. Secretary of Energy Dr. Steven Chu has indicated his support for technological solutions to climate change, including "benign" geoengineering schemes such as roof whitening.[13] An odd effect of geoengineering's mainstreaming has been an alignment of the positions of some interest groups that were previously diametrically opposed. While some longtime climate scientists such as Paul Crutzen and Ken Caldeira claim to have only gradually and reluctantly embraced geoengineering, fearing devastating effects from climate change, a new and powerful corporate lobby for geoengineering has emerged in recent years made up of people whose motivation has never been concern for the environment or the world's poorest people.

In 2008, Newt Gingrich, former Speaker of the House of Representatives and a senior fellow of the American Enterprise Institute—a neoconservative think tank promoting free enterprise—sent a letter to hundreds of thousands of Americans urging them to oppose proposed legislation to address global warming. "Geoengineering holds forth the promise of addressing global warming concerns for just a few billion dollars a year," wrote Gingrich. "Instead of penalizing ordinary Americans, we would have an option to address global warming by rewarding scientific innovation.... Bring on the American Ingenuity. Stop the green pig."[14] For those who previously doubted (or still do) the science of anthropogenic global warming, the geoengineering approach shifts the discussion from reducing emissions to an end-of-pipe solution. Once geoengineering is an option, there is no longer a need to bicker about who put the carbon dioxide in the atmosphere (or ask them to stop). If we have the means to suck up greenhouse gases or turn down the thermostat, emitters can continue unabated. At least one commentator has charged that the wholesale embrace of geoengineering by industry-friendly think tanks represents a deliberate tactic of distraction and delay by the same folks who use oil company dollars to discredit the science of climate change. "If we can be made to believe that mega-scale geoengineering can stop climate change, then delay begins to look not like the dangerous folly it actually is, but a sensible prudence," explains Alex Steffen, editor of *Worldchanging: A User's Guide for the 21st Century.*[15]

Unfortunately, humanity has already proven massive Earth restructuring to be wonderfully operational. Fill wetlands, dam rivers, introduce crop monocultures, wipe out species—and the ecosystem changes. Cut down forests and pump massive amounts of pollution into the atmosphere—and the climate changes. While these old-fashioned ways to geoengineer the planet are well known, the newer technologies are less familiar.

They include proposals to create vast monoculture tree plantations for biochar, biofuels production, and carbon sequestration; "fertilize" the ocean with iron nanoparticles to increase carbon dioxide–sequestering phytoplankton; build trillions of space sunshades to deflect sunlight 1.5 million kilometers from Earth; launch 5,000–30,000 ships with turbines to propel salt spray into the air to whiten clouds to deflect sunlight; drop limestone into the ocean to change its acidity so that it can soak up extra carbon dioxide; store compressed carbon dioxide in abandoned mines and active oil wells; blast sulfate-based aerosols into the stratosphere to deflect sunlight; and cover deserts with white plastic to reflect sunlight.

While the following list is not comprehensive, there are compelling reasons to say *no* to geoengineering:

- Too-perfect excuse: Geoengineering offers governments an option other than reducing greenhouse gas emissions. For many industrial advocates, geoengineering means "buying time"[16] to avoid action on emissions reduction.
- Large-scale: For any geoengineering technique to have a noticeable impact on the climate, it will have to be deployed on a massive scale, and any unintended consequences are also likely to be massive. We don't know how to recall a planetary-scale technology.
- Unequal: OECD governments and powerful corporations—which are responsible for 90 percent of historic emissions –are the ones with the budgets and technology to execute geoengineering's

gamble with the planet. There is no reason to trust they will have the rights of more vulnerable states or peoples in mind.

- Unilateral: Many geoengineering techniques could be relatively simple to deploy, and the technical capacity to do so could be in some hands (of individuals, corporations, states) within the next ten years. It is urgent to develop a multilateral mechanism to govern geoengineering, including establishing a ban on unilateral attempts at climate modification.

- Unreliable: Geoengineered interventions could easily have unpredicted consequences due to mechanical failure, human error, inadequate understanding of Earth's climate, future natural phenomena (such as storms or volcanic eruptions), transboundary impacts, irreversibility, or funding failures.

- Treaty violation: Many geoengineering techniques are "dual use" (i.e., have military applications). Any deployment of geoengineering by a single state could be a threat to neighboring countries and, very likely, the entire international community. As such, deployment could violate the U.N. Environmental Modification Treaty of 1978, which prohibits the hostile use of environmental modification.

- Commercializing the climate: Competition is already stiff in the patent offices among those who think they have a planetary fix for the climate crisis. If geoengineering's "Plan B" were ever put into motion, the prospect of it being monopolized is terrifying.

- Carbon profiteering: No commercial interests should be allowed to influence the research and development of such serious planet-altering technologies. If, as advocates insist, geoengineering is actually a "Plan B" to be used only in a climate emergency, then it should be forbidden to be considered for carbon credits or to be used to meet emissions-reduction targets.

Ultimately, humility will need to replace hubris if humanity is to effectively address the predicament that climate changes poses to the future of life on Earth. The first steps toward some control on these technologies was taken when the 193 parties to the Convention on Biological Diversity adopted a de facto moratorium on geoengineering activities in Nagoya, Japan, in October 2010. The challenge now is to enforce the moratorium on all real-world geoengineering experiments and ultimately convert it into a strong test-ban treaty that would prohibit any unilateral attempts to geoengineer the planet. No patents should be granted to geoengineering technologies, which would be a perverse incentive to move forward with these dangerous schemes. Private sector involvement in experimentation or deployment of geoengineering technologies should be prohibited. And no offsets or carbon credits should be allowed for geoengineering technologies. It's time for a host of ecologically and socially restorative actions to stem the climate crisis, not a technologically risky attempt at global climate control.

Part Five

UNDER ATTACK

INTRODUCTION

Onslaught of the Energy Machine

The late Garrett Hardin wrote, "Loss of freedom is an inevitable consequence of unlimited population growth in a limited space." The Earth is most definitely a finite place, and thus the human population explosion, with every upward demographic tick, represents an incremental diminishment of liberty for people and wild nature.

This incrementalism may be one reason why we are slow to recognize the cumulative effects of the energy machine. No one sees the entire system, and in fact, much of the system is designed to be out of view. But every new gas or oil well, every new power plant and hydroelectric dam, every additional pipeline and tanker ship, every solar panel and wind turbine has ecological costs.

Ecosystems, of course, are not eliminated in a day. Longleaf pine forests and tallgrass prairie are both more than 98 percent diminished from their extent prior to European settlement in North America; that loss took place over centuries through the unwitting actions of innumerable individuals who cleared the forest and plowed the prairie. With humanity now surpassing 7 billion people, the phenomenon of nature dying a death by a thousand cuts is magnified—it is billions of little cuts whittling away at the integrity of natural habitats and processes.

After agriculture, the aggregate impacts of energy production, transmission, and use are perhaps the most significant mark that humans make upon the biosphere. The residue of the energy machine is everywhere, from the radioactive isotopes in the breast milk of Inuit women in the High Arctic to the coal dust in the lungs of Australian miners. The chemistry of the atmosphere records our exploitation of energy resources along with the associated destruction of carbon-storing forests, grasslands, and soils. With their loss go species, headed into the endless night of extinction.

Certainly no coal miner or pipeline builder wakes up in the morning and thinks, "Let's go kill some wildlife today." Habitat loss and degradation are merely unintended side effects of most energy development, and many people who work in the energy business are actively working to lessen environmental impacts. But like those earlier loggers and farmers who cut down the piney woods or converted wild prairies into cornfields, individuals may be well-meaning even while a culture's overall effect on the natural world is negative. If the worldview driving the system is anthropocentrism—that is, the Earth is merely a storehouse of resources for our use and profit—then the result can hardly be otherwise.

We cannot predict every contour of the future but we can certainly recognize the particulars of the present and see that the path we are on is toward more domestication of nature, more loss of beauty, more extinction. Not, perhaps, of *Homo sapiens*, but of myriad species with which we share the planet. The onslaught of the energy machine does not just threaten the lungs of asthmatic children downwind of coal-burning power plants or the livelihoods of farmers around the Fukushima Daichi nuclear station, it threatens the whole—a living system that had perfected an elegant energy economy for billions of years before we even arrived.

WILL DRILLING SPELL THE END OF A QUINTESSENTIAL AMERICAN LANDSCAPE?

ERIK MOLVAR

THE NATURAL RESOURCE *perhaps most important to the character of the West is the region's wide-open spaces, but that sense of remoteness is being eroded by the bulldozers and drilling rigs sweeping across the region. Oil and gas development poses serious threats to air and water quality, fragments wildlife habitat, and generally degrades the health of the land. Public opposition to the West's energy boom is growing as spectacular landscapes are degraded.*

❖

From the windswept plains of Montana to the deserts along the southern border of New Mexico, a battle is raging between the fossil fuel boosters and westerners fighting for their sense of place. At stake is the very soul of the American West. The wide-open spaces, the spectacular landscapes, the sense of remoteness are all being eroded by the bulldozers and drilling rigs sweeping across the region. And the oil and gas industry, once viewed with patriotic fervor, has become the object of concern and distrust across the West. Oil and gas development poses serious threats to air and water quality, can destroy wildlife habitat, and generally degrades the health of the land.

AN ASSAULT ON OPEN SPACE

The natural resource perhaps most important to the character of the West, yet given least weight in land-use planning, is the region's wide-open spaces. For many years, a prevailing frontier mentality has clung to the myth that open spaces are inexhaustible, that because spectacular landscapes shaped by nature have been abundant here since time immemorial they will never disappear. But as public lands and private ranches have been converted to industrial landscapes by oil, gas, and coalbed methane

drilling, in increments ranging from thousands to millions of acres, westerners have been confronted by the reality that, while open space may be
considered a birthright, it is not limitless.

Public backlash against drilling has grown especially fierce in cases
where oil and gas development has invaded special landscapes having
elevated value for westerners. In Montana, for example, proposals to
drill exploratory wells in the mountains of the Rocky Mountain Front
prompted decades-long efforts to close the area to industrial incursions.
In Wyoming, threats of drilling in high-value landscapes in the Red
Desert, such as Adobe Town and the Jack Morrow Hills, caused the
largest outpourings of public opposition to development in the state's
history, while a proposal to site exploratory wells near Little Mountain,
one of Wyoming's premier elk hunting destinations, galvanized the local
hunting community to fight the project. In Colorado, the prospect of
drilling rigs moving into proposed wilderness atop the Roan Plateau
sparked a major controversy, and in southern New Mexico concerns
about protecting an important underground aquifer spurred then governor Bill Richardson to file a lawsuit to protect Otero Mesa. All of these
examples have this in common: Human residents of the affected states
saw outstanding benefits in these landscapes that trumped the value of
any oil and gas that might lie beneath them.

The sacrifice of some of America's last wild and natural landscapes
comes with long-term costs to western economies. The tax rolls swell
during oil and gas booms, but during the busts that inevitably follow,
tourism and other industries dependent on open spaces serve as the
sustainable economic engines that keep western economies afloat. In
Wyoming, tourism is the second-largest economic sector, with hunting,
fishing, and wildlife watching on public lands contributing more than
$100 million to the Wyoming economy. Protecting open space from the
depredations of drilling makes sense: A study examining the economies
of rural communities throughout the West found that communities with
a larger percentage of public lands in protected status (such as national
parks, wilderness, and roadless areas) had greater population growth,
more job creation, and higher growth rates for personal income; unprotected federal lands showed no such positive economic correlation. And
while oil corporations based in Texas and Oklahoma contribute significantly to the local tax base during the boom years, most of the mineral

wealth they generate leaves the region without triggering any economic multipliers there.

THE DESTRUCTION OF THE WEST'S WILDLIFE HERITAGE

While sagebrush ecosystems in the Great Basin states of Oregon, Nevada, Utah, and Idaho have been badly impaired by the invasion of cheatgrass, a noxious weed, the sagebrush ecosystems of Wyoming and northwestern Colorado remain relatively intact. This vast sagebrush steppe is among the healthiest natural ecosystems remaining on the North American continent. Unfortunately, lying beneath the remote sagebrush basins and surrounding mountain ranges of the Rocky Mountain states are lucrative deposits of natural gas and coalbed methane, along with a marginal amount of oil. The collision of big oil and the nation's finest wildlife resources and open spaces threatens some of America's wildest remaining country, and has ignited a controversy pitting westerners of all types against oil executives.

The impacts of drilling activity on wildlife have been well documented, including the direct elimination of habitat as thousands of acres fall beneath the bulldozer blade, increased poaching as temporary workers from out of state flood the gas patch, and a rise in wildlife deaths due to vehicle collisions resulting from surging truck traffic. Noise from drilling rigs and compressor stations can disturb wildlife enough to cause temporary or long-term abandonment of otherwise suitable habitats by some species.

But the biggest impact on wildlife of the oil industry's incursion into the intermountain West is habitat fragmentation. Imagine a vast and untouched swath of sagebrush covering hundreds of square miles. Then build a road or establish an oil or gas well pad right through the middle of it, either of which will be avoided by wildlife. The habitat has just been carved into two large fragments. Continue to divide these fragments into smaller and smaller parcels until there is a road or a well pad every quarter mile. Now you have the equivalent of full-field development for natural gas. The remaining tatters of natural landscape are often too small to provide sufficient habitat for many types of wildlife. When this occurs, species sensitive to habitat fragmentation and disturbance by human activity—ranging from the pygmy rabbit to the ferruginous hawk to elk—are the first to disappear.

The oil industry is fond of claiming that even a fully developed well field disturbs less than 5 percent of the landscape, leaving the vast majority as untouched habitat. But the fact of the matter is that while only a small proportion of this land may fall directly under the bulldozer's blade, the impacts of production activity radiate outward from roads and well pads to the greater surrounding area, driving away sensitive wildlife. Populations of sagebrush songbirds decrease within 300 feet (91 meters) of well-field roads. Elk avoid lands within a half mile of well-field roads in sagebrush country. Sage grouse are even more sensitive: Drilling rigs sited within three miles of sage grouse breeding sites (called "leks") caused declines in breeding birds in a western Wyoming study, and even after drilling was finished, a producing well sited within two miles of a sage grouse lek caused populations to drop significantly. The dust thrown up on gravel roads by regular truck traffic can choke off vegetation: For one coalbed methane project in the eastern Red Desert, the Bureau of Land Management predicted that the amount of usable forage for livestock and wildlife could drop by almost a third as a result of dust. As a result, the typical oil or natural gas field with four wells per square mile might have only 3 percent of its land area occupied by roads, well pads, and pipelines, but—from a wildlife perspective—it becomes an industrial landscape with fragmented habitats vacant of sensitive native wildlife.

Despite industry claims that wildlife can happily coexist with drilling rigs in the gas patch, in reality some species dwindle as the drilling rigs move in. Coalbed methane development in the Powder River Basin caused sage grouse populations to decline by 82 percent (in contrast, sage grouse in undeveloped areas dropped by only 12 percent), while population modeling in western Wyoming showed that major declines in the Pinedale Anticline and Jonah Field would lead to the total loss of sage grouse there within nineteen years. Another study showed that mule deer populations wintering in the Pinedale Anticline drilling area dropped by 43 percent in conjunction with gas development, while nearby herds on ranges free from drilling did not exhibit similar declines.[1] In addition, oil and gas fields can be an obstacle to big game migration. There is evidence that elk migrations subjected to oil development in the LaBarge Field at the base of the Wyoming Range have been blocked from accessing winter ranges north of LaBarge Creek by the development of an oil field at the base of the foothills.[2]

SOLUTIONS ARE AVAILABLE BUT PROGRESS REMAINS ELUSIVE

Every drop of oil and every cubic foot of gas extracted was carbon already sequestered underground, which oil corporations are pulling out for the express purpose of burning, with the carbon going straight into the atmosphere. Natural gas production comes with added problems: Methane, which makes up over 85 percent of natural gas, is a far more potent greenhouse gas than carbon dioxide, and massive amounts of it leak out at well sites, compressor stations, and along pipelines. In addition, production of coalbed methane from near-surface coal deposits can result in methane seeps at the surface that vent thousands of cubic feet per minute into the air.[3] Once the water is removed from the coal seam to release the coalbed methane, ideally the gas is captured by production wells and put into pipelines for sale—but significant quantities can escape to the surface through fractures in the bedrock or at locations where the coal seam crops out at the surface.[4] Thus, from a global climate change perspective, the ultimate solution might be to simply wean our economy off its addiction to fossil fuels.

However, given the time it could take for renewable sources of energy to replace oil and gas, it is unlikely that drilling for oil and gas will stop completely in the next several decades. Even assuming that full-field development of major oil and gas deposits will continue, such development could become compatible with protecting wildlife and treasured landscapes if strong reforms of drilling management are imposed on the oil industry. The first and most obvious change would be to recognize that the highest value inherent in at least some of our public lands is their wildlife, recreational, or scenic attributes, not drilling potential; these lands should be removed from industrial development.

The second major change would be to require that full-field drilling operations be designed to maximize the ability for wildlife and native ecosystems to survive alongside drilling by minimizing the physical footprint and other impacts on the lands, wildlife, clean air, and water resources.

In some parts of the world, directional drilling is used to tap oil and gas deposits that are as far away as seven horizontal miles from the well site. Additionally, this method allows for clustering more than 50 individual gas wells at a single pad and drilling outward to tap surrounding lands, instead of building a road, pad, and pipeline for each individual well. Directional drilling also makes it possible to drill diagonally underneath sensitive wildlife habitats, leaving the surface of the land undisturbed. These methods cost a little more than the heavy-footprint methods of designing well fields that are typical today, but if every company were held to the same high standard, it might inspire the industry to apply some American ingenuity so that very little oil or gas would be unavailable for production in the end.

Finally, there is no reason why all federal lands should be open to drilling at any one time. It might be wiser to try "phased development," in which a fraction of land is open for drilling at a given time and later, after the industry has left and the land has been reclaimed, a new area could be opened up. Wildlife populations would always have at least some habitat in which to survive, and regional economies would be supported by a steady pace of industrial development instead of today's devastating cycle of boom and bust—a cycle that stretches communities to the breaking point as oil-field workers flood in, and then leaves them high and dry with no economic base to support the expanded infrastructure that was built to support the boom in the first place.

Although far from being a perfect system, on U.S. Forest Service lands, federal foresters decide which lands will be offered up for logging, design the size and layout of different cut units, and even determine which trees will be cut and which must be left standing. In contrast, oil and gas development on public lands has been managed—from cradle to grave—almost exclusively by the fossil fuels industry itself. The public lands offered for lease at oil and gas auctions are identified by industry, not by land managers. Once they own the leases, the industry decides where and when to drill exploratory wells and conduct seismic testing. When it comes time to design where and how many wells to drill in an oil or gas field, it is the corporation that designs every aspect of the well field; federal officials can influence the design, but typically the alternative backed by the proposing corporation is accepted, with few or no modifications. Nothing in the law, however, prevents federal officials in the Bureau of

Land Management and other agencies from taking the reins to manage oil and gas drilling in a way that makes it ecologically sustainable. It's high time they did.

For years, the oil industry has been able to pursue maximum profits at the expense of respect for the land. That they would is no surprise: Private corporations are accountable to their shareholders, and quarterly profit margins are the principal measure of success. But across the American West, more than half of the land and an even greater fraction of the oil and gas deposits are under federal ownership, which means that they belong to the American people. And these lands are supposed to be managed for multiple purposes, including wildlife, clean water, wilderness, and public recreation. There is no particular reason that oil and gas drilling should be the dominant use of the land as it is today, with other uses surviving as best they can in the context of drilling every profitable deposit of oil or gas. Why shouldn't the needs and priorities of the public come before private profits when it comes to our public lands?

BACKING THE FRONT

Fighting Oil and Gas Development in Montana's Rocky Mountain Front

GLORIA FLORA

WHEN GLORIA FLORA *took the helm of Lewis & Clark National Forest in Montana in 1995, she found priceless wildlands threatened by oil and gas speculators. Defying convention, she declared the area off-limits to oil and gas development, adding a definitive new twist to the interplay between community groups, the fossil fuel industry, and the government that is playing out in surprising ways.*

❖

Squinting out over the chatting hikers, I stared in amazement at our stunning backdrop, Montana's Rocky Mountain Front. We were going to explore one of its ridges today under the guidance of Lou Bruno from the Montana Wilderness Association. I had just arrived in Montana a few days earlier to literally get my feet on the ground before starting my new job as Forest Supervisor of the Lewis & Clark National Forest. Taking advantage of one of the Association's Wilderness Walks would connect me with a part of my new forest and some citizens who cared about it. My first footsteps in 1995 into those mountains became my first footsteps into decades of working to protect an incredible landscape.

The Rocky Mountain Front in Montana, or as it's known locally, "the Front," runs from the Canadian border almost to Helena, about 150 miles south. This is the Overthrust Belt, the eastern edge of the Rocky Mountains which have literally smashed into the Great Plains, forcing thick strata of limestone 4,000 feet in the air above the rolling grassy plains. Row after row of these reefs jut east, backed by the Bob Marshall, Scapegoat, and Great Bear Wildernesses. These formations have created diverse and biologically rich ecosystems that teem with wildlife, fish, and precious water. Every species that Meriwether Lewis and William Clark encountered on their 1804–1806 journey of discovery still thrives here, with the unfortunate exception of free-roaming bison. Only superlatives can describe the Front, home to the second largest elk herd in the United States, the largest wintering herd of mule deer, and every charismatic carnivore you find in the West. Threatened and endangered species abound and are rebounding: Grizzly bears here are becoming plains animals again, as they were two hundred years ago.

Over 365,000 acres of this intriguing landscape are sandwiched in between the wilderness and the sparsely populated private ranch country. This public land is managed by the U.S. Forest Service, with some Bureau of Land Management (BLM) lands scattered along its edge. Despite over one hundred years of management, the Forest Service is a latecomer after the Blackfeet who call this country *Miistakis*, the "Backbone of the World" where they were created. Then came the explorers, trappers, miners, and ranchers carving their existence from this beautiful but unforgiving landscape. After the Forest Service came the outfitters, the loggers, and the oil companies. Others showed up as well, in the guise of "normal folks," people from all walks of life who were either drawn like a magnet to this landscape or were born here and just couldn't bring themselves to leave. And they started forming a quiet line of resistance against the unraveling of the Front.

GAS OR GASSED?

An eighteen-year Forest Service veteran, I took the helm at Lewis & Clark National Forest—which includes much of the Rocky Mountain Front—just shy of my 40th birthday. I soon learned that much of the Front had been leased to oil and gas companies fantasizing about the possibility of large pockets of gas trapped beneath its formidable surface—and not without reason, as Canada's Rocky Mountain Front in Alberta is heavily veined with roads, pipelines, well pads, and all the appurtenances of gas depletion. (The common use of the term 'gas production' rankles me. We're not 'producing' fossil fuels, we're depleting something produced millions of years ago.) A few exploratory wells from decades past gave some excitement but nothing commercially viable. Many geologists believe the layers that hold the gas on the Canadian Front become progressively thinner—down to as little as a few feet— this far south. Some spectacularly deep holes drilled in recent years have come up dry.

My staff at Lewis & Clark was in the throes of an oil and gas leasing analysis when I arrived. Just five years before, a new law shifted responsibility

quietly explained the conclusions reached from the extensive scientific
and cultural information in the environmental impact statement (EIS).
Ecologically and culturally the Front was too important to develop.
People's sense of place, their attachment to this land, and the values of its
wildness were worth more than energy development.

The response still moves me. I received over four hundred letters and
phone calls from around the country thanking me. I got notes from Forest
Service employees I didn't know saying, "Today I was proud to put on my
uniform." People sent me flowers and even pictures of their grandchildren
saying, "Here's who you made that decision for."

A surprisingly small segment of the upset fossil fuel industry took
the decision to court. In an attempt to soften public opinion, they tried a
variety of tacks. Perhaps as a woman, she was too soft-hearted to the pleas
of citizens, or maybe she was premenopausal and not thinking clearly?
Maybe she made this whole sense-of-place thing up? One industry rep
summed up the problem (and I swear to God she actually said it), "She
listened too much to the public." By this time, my agency, seeing the flood
of positive support, boldly stepped forward taking bows.

The judges at the District and the 9th Circuit Courts couldn't find a
single point in industry's arguments that bore support and the case held
completely in favor of the Forest Service. The plaintiffs didn't stop until
the Supreme Court refused to hear the case.

All victories are but steps along the way. In 1998, I went off to the
Humboldt-Toiyabe National Forest in Nevada, where I ended up resign-
ing from the Forest Service to call attention to the anti-federal government forces who were behaving in abysmal and life-threatening ways to
the land and to my employees. They were egged on by a few right-wing
members of Congress with a mission to emasculate the Forest Service and
its regulations. But that is a story for another day.

THE COALITION IS FORGED

Now free to choose where to live and what to do, I returned to Montana,
to the Front. I started Sustainable Obtainable Solutions (SOS), a nonprofit
organization dedicated to ensuring the sustainability of public lands and of
the plant, animal, and human communities that depend on them. It's the
Forest Service mandate in my words.

I circled back to the Front like a hummingbird honing in on a bedazzling red flower, There of course had been no new leases, but interest began to stir in drilling on leases just south of Glacier National Park bordering the Blackfeet Reservation in an area steeped in rich tradition and history known as the Badger-Two Medicine and in the Blindhorse area just north of the Teton River.

It felt like a David-and-Goliath matchup now with George W. Bush and Dick Cheney in the White House, a scion of oil and a former oil company CEO. What were the odds we could keep their pals at bay, even on the Front? The individual efforts of concerned citizens and regional organizations seemed no match for the fossil fuel industry juggernaut. A call from Gene Sentz, schoolteacher, part-time guide, and full-time engaged citizen, shifted gears for all of us. Gene had started Friends of the Rocky Mountain Front in the 1980s to bring together locals who didn't want to see the Bob Marshall Wilderness carved up with mines, seismic lines, and drill pads·sprinkling the Front like toxic confetti. (They deserve a round of applause, along with Congressman Pat Williams, for stopping the wilderness seismic work with the Don't Bomb the Bob campaign.)

Gene suggested that his and other groups join up, combining our resources and brains to bird-dog issues on the Front. Four organizations joined with the Friends and became the core of the Coalition to Protect the Rocky Mountain Front, or CPRMF: Montana Wildlife Federation; The Wilderness Society/Northern Region; Montana Wilderness Association; and my group, Sustainable Obtainable Solutions. "United we stand" proved very fruitful.

We purposefully opted not to become an official nonprofit. Members were simply those who said they were members. No hierarchy, no designated leader, no bylaws—then or now. We had the power of a single purpose: *To protect and defend the biodiversity, beauty, integrity and stability of the Montana Rocky Mountain Front as it has flourished through the ages.* It bound us more solidly than any business model. We applied for grants jointly, designating one of the core bona fide nonprofits to share and manage the grant.

A plethora of organizations, citizens, and businesses joined in our meetings and conference calls. We added detail to the mission with a shared set of principles. Then we set to work with a defined strategy, timelines, and responsibilities. Each member's skills were assessed and roles assigned. SOS became the tribal/public lands liaison and national media coordinator. Our weekly organizers' calls began—and continue now, a decade later.

INROADS FOR NO ROADS

In 2001, the Lewis & Clark National Forest was struggling to update their Travel Plan (the plan that determines which roads and trails will be open, and when, to motorized or nonmotorized travel, for the next fifteen to twenty years). Given the high stakes placed by people on the freedom to recreate how and where they want, this access planning is arguably one of the most contentious National Forest processes. Under the old Travel Plan in sensitive areas, like the Badger-Two Medicine, ATVs and motorbikes had carved deep eroding channels across steep slopes. And dirt bikes were now scaring the bejesus out of horses sharing the trail.

The Forest tried to shortcut the planning process by developing a disastrous in-house revision that greatly expanded motorized use. In short order, 2,000 letters of protest, generated by the Coalition, forced the Forest Service back to the drawing board.

In the meantime, energy leasing and development were running amok on public lands throughout the West, with corporations taking advantage of the leniency, encouragement, and perverse incentives of the Bush administration. Agencies like the BLM were directed to streamline permit processes, setting up entire field offices devoted to rubber-stamping energy permits with performance rated on the number of permits completed. On the ground, inspection and enforcement were light to nonexistent, with reprimands for those who discovered too many problems or slowed the process.

In 2001, Startech/Thunder Energy, a Canadian company, proposed exploratory drilling for deep-well natural gas in the BLM's Blindhorse Outstanding Natural Area on the Front. Yes, you read that correctly. BLM rules allowed for oil and gas drilling just about anywhere including these outstanding areas. Perched high on a verdant plateau overlooking the plains, the proposed drill site would be accessed by a road requiring four switchbacks across the mountain face. Over one hundred semitruckloads of materials would be hauled up to drill four horizontal wells in the unyielding rock toward targets as small as a few feet in diameter, testing the limits in distance and accuracy of the nascent technology.

Concurrently, the Arctic National Wildlife Refuge (ANWR) was also in the crosshairs of the oil industry and the Bush administration. Both the Front and ANWR exemplified the finest gems in our national treasure chest of public lands. We knew if the Bush administration could break the back of the public and of the environmental movement by drilling and roading these two places, then no place was too precious to drill.

The Coalition developed a comprehensive strategy to expose this ill-conceived desecration of the Front for what it was—destruction of the beauty and ecological integrity of the land, stripping future generations of irreplaceable values. This wasn't just about saving the Front anymore; this was about defending the American legacy of wildlands. We were determined to show, using the Front as the exemplar, that when the people lead we can make the leaders follow.

SEEING IS BELIEVING

Armed with clear, compelling information and photographs, we stayed on message and we kept delivering it. The Coalition was completely clear that our intent was to stop this proposed development and our reasons were sound. The proposed drilling and what it could lead to in a full-field development scenario would destroy irreplaceable values. Wildlife, including threatened and endangered species such as the grizzly bear, wolf, and cutthroat trout, as well as the abundant deer, elk, mountain goat, and bighorn sheep populations would be severely impacted. Cultural values such as the traces of the old North Trail, over ten thousand years old, were threatened. Water, the lifeblood of the communities, ranches, and farms, would be impacted. Historical and traditional uses and values would be laid to waste. All for what geologists estimated was a 2–3 percent chance of finding an economically viable gas field.

Following the adage "Seeing is believing," we partnered with Bruce Gordon of EcoFlight, an organization specializing in providing an aerial perspective on threats to the environment. Over the past decade, through EcoFlight, the Coalition has arranged air tours of the Front for reporters, photographers, opinion leaders, and public officials. With a local Coalition member as guide, the flights cruise up the Front with its jaw-dropping beauty and untrammeled sweeps of land worthy of national park designation. The fact that this incredible land lacks any particular designation or protection starts to sink in. How did this land escape notice? And how has it remained undeveloped for so long?

We explain the hundred-plus years of conservation history of the Front: a grassroots-driven (but not particularly coordinated) saving of one piece at a time. Over the Blindhorse Outstanding Natural Area, the site of the proposed drilling, we circle. "You've got to be kidding," is frequently heard at this point.

We glide north past Glacier and Waterton National Parks into the gas fields of Alberta. Here, the land—or what's left of it—speaks for itself. Overlain on the same once-stunning landscape as the Front are the trappings and appurtenances of full-on energy development. It's heartbreaking. Every drainage has a road up the bottom, each with multiple drill pads. Clouds of dust rise from the heavy truck traffic, blending with the pyre-like fires of flaring gas. (On the ground, the deep whooshing sound of combustion is only overridden by deafening compressors.) Pipelines spread through forest and farms. Taking advantage of the openings, erosion from ATV tracks follows the pipelines, just as the timber companies have followed the roads. Clear-cuts dangle down steep mountainsides above the drill pads. I recall Governor Brian Schweitzer, a candidate at that time, shaking his head disbelievingly. "They said the footprint would be small."

And then there's the sweetening plant. There's no need to say more, so we just turn around here over the sprawling industrial plant taking up over a hundred acres with its trailing network of railroads and roads. The sour gas in this area is laden with sulfur dioxide, a poisonous gas that needs to be removed (hence, the gas is "sweetened"). The by-product is sulfur—a lot of it. The neat geometric piles of bright yellow sulfur look big enough from the air, but on the ground they dwarf you. Multiple football fields in size, they tower three stories in the air, with pale yellow clouds of dust wafting off them. Local ranchers complain of the high rate of cattle mortality here. The gas is sour in Montana too.

The Coalition's first billboard juxtaposed two photographs, one of the Front in all its glory and the other of the Alberta sweetening plant. Above them, respectively, "A long-term love affair?" and "Or a one-night stand?" It got your attention.

Momentum continued to build in our favor but every step was

hard-won. Veteran county commissioners, who never saw a development they didn't love, stumped hard for industry. Attempts were made to undermine the Coalition, claiming we were national groups trying to manipulate local opinion. But by this time, local citizen members were deeply engaged with the Coalition and were calling the shots as often as the original coordinating organizations. The public, becoming better informed, started to ask hard questions and express strong opinions against drilling.

A quiet parallel effort outside the Coalition began gathering names of people who would be willing to protest physically, putting themselves between the bulldozers and the Front if it came to that. The list quickly swelled to over four hundred names before we stopped counting. This was our public land and our grandchildren's public land and we weren't going to sell out.

THE ARMOR CRACKS

The Front and its spokespeople appeared in documentaries and major media, from the *Los Angeles Times* to the *New York Times*, from *Time* magazine to *The Economist*. Editorials urged protection and reports chronicled how far the Bush administration was willing to go to feed the hungry maw of energy companies. PBS's Bill Moyers sent David Brancaccio out to cover the story. David and producer Brian Myers were spellbound by the landscape and its stories and the fate that could befall it. The fifteen-minute piece they produced for 'NOW with Bill Moyers' tracked my history with the Front and its current peril.

The day before the show's airing in October 2004, the producer got a call from the Department of the Interior's Press Office. "We know you are going to excoriate us. We want to inform you we are terminating the EIS." (That is, they were canceling the project.) "You need to pull the piece or announce our intention." When Brian relayed that to me, I said, "They're lying. Ask them…" and I proceeded with a list of questions and requests. The call came back; they had answered all the questions. I still didn't believe them and sent back another set of questions which they answered satisfactorily. The piece aired with the closing note that the EIS had just been withdrawn. It was the first and only time an oil and gas EIS had been terminated midstream.

The Department of the Interior formally announced not only that the BLM was terminating the EIS on the proposed drilling but also that they were advising current leaseholders to consider selling their leases back to the government and were instituting programs to protect up to 170,000 acres of private land by buying conservation easements from willing sellers. You could have knocked me over with a feather.

I'd love to give the Coalition full credit for this, but we were just luckily in the right place on the timeline of history: one month before elections and Bush's reelection bid. He had announced to conservative sporting groups, after decimating wetlands protections, that indeed there were places too special to drill (although he failed to identify any). Likely his administration was flailing around for a place to make good on his statement. And here was the Coalition making an unending ruckus in the media about the Front, scoring points by describing what Bush's ruthless energy policy looked like on the ground. And Thunder Energy was a Canadian company that obviously had not contributed directly to Bush's campaign. Since ANWR wasn't negotiable, suddenly the Front looked like just the right bone to throw. Call me cynical. Maybe it was just divine intervention.

It turns out BLM's Montana Director Marty Ott had informed his supervisors of the unyielding opposition and the growing costs of the proposed drilling project; the BLM had already spent over $1 million and was just getting started. Ott had also described to the proponents the real timeline—under the best circumstances they could be looking at a decade before a bit might bite the Front's soil. The Coalition had many tools as well as overwhelming public opinion in our court, and Ott knew we were more than ready for engagement in an actual court.

Jubilant from our victory of stopping the proposed drilling, we turned our attention back to gaining more permanent protection as my original 1997 moratorium was running out. Sending Coalition members back to Washington, D.C., to meet with our representatives and agency heads proved very effective. And it was pure West: cowboys and American Indians, with the right boots and hats—and many of them Republicans. Despite then-Senator Burns (R–MT) having declared publically that I was an idiot who never made one right decision, in October of 2006 he introduced legislation to make my moratorium permanent and also ban hard rock mining. Burns didn't survive reelection so Senator Baucus (D-MT) picked it up and helped it pass into law by January 2007.

We had a big public party for that one!

WE'RE ON A ROLL

The Forest Service released their new draft Travel Plan in 2005. The Coalition helped generate 37,000 comments, the vast majority supporting quiet recreation and wildlife protection. The final Travel Plan released in 2007 was reasonable, rational, and ground-truthed. With few exceptions, it fit traditional patterns and showed true regard for the wildlife and fish of the Front. In the Badger-Two Medicine, thanks to Keith Tatsey and other leaders in the Blackfeet Nation, the Forest Service honored the no-motorized-traffic alternative supported by the Blackfeet. Score again!

Now we were ready to move forward with the Big Idea: a citizen-driven collaborative effort to design a bill that would indeed permanently protect the Front and keep it like it is. This was shaky ground for us organization types accustomed to being campaign leaders and the designers of bills. Our plan was straightforward: We'd work in concentric circles of stakeholders. We'd kick off the first meeting, but then the inner circle, the Coalition members, had to hammer out a vision and strategy, form a Steering Committee, and hone the tactics. We aimed for consensus, even if it was given grudgingly.

The next circle included more of the community, especially those with a vested interest in keeping the Front the way it is, including people we assumed would be opposed. The proposal was certainly going to add wilderness, include grazing leases, and put the fear of uncontrolled wildfire into people's hearts. Those are all fightin' words in the West.

It was pretty brutal at times. Negotiations were hard-won and the lines on the map moved regularly. Feelings were hurt, and some people walked out (they usually came back). But we pressed on. For four years, we held planning meetings, personal visits, surveys, interviews, barbeques, public meetings, agency meetings, tribal meetings, events, and traveling lectures, and we published press releases, articles, op-eds, editorial board visits, economic studies, and even a visitor's map in partnership with the Chamber of Commerce that rivals a national park brochure. Coalition members went to scores of organization meetings to explain what we were building and to ask for input.

We worked closely with state and federal agency people and politicians. We gathered supporting organizations (numbering 65 and rising), opinion leaders, and elected officials. Three former Chiefs of the Forest Service and five former Forest Supervisors of the Lewis & Clark National Forest signed on.

The weekly conference calls continued. Jennifer Ferenstein of the Wilderness Society in Montana rode herd and kept us organizers on point with remarkable aplomb and effectiveness. The organizers' faces have changed over time but the intent and commitment has never wavered. I have deep respect for every Coalition organizer and member of the Steering Committee, each driven by a selfless passion for the Front. They have given me and the American public a gift beyond measure.

From the chaos emerged The Rocky Mountain Front Heritage Act, a fine citizen-crafted, highly supported act that does three things: adds lands that are already managed as de facto wilderness to the actual Wilderness system; designates the rest of the lands as a Conservation Management Area; and raises the priority of fighting noxious weeds on public and private lands. After answering some hard questions and satisfying his concerns, Senator Baucus agreed to sponsor the bill. In March of 2011 the bill was heard in the Energy and Natural Resources Committee. As I write we're hoping the Rocky Mountain Front Heritage Act comes before Congress in the next few months.

But dark clouds are on the horizon. The latest threat is increased oil and gas leasing and exploration on private and state lands along the Front. Some believe the Front may be the new Bakken Formation, which has fueled an energy boom in North Dakota. Using new technological developments like horizontal drilling and fracking, energy companies believe they can finally make a killing on the Front. Just in the past year hundreds of thousands of acres have been leased by oil companies along the Front, and exploratory drilling is now occurring. The worry of some is that if the Heritage Act isn't passed soon, the momentum for energy development and perhaps even a reversal on the drill ban for the federal lands will make the public lands of the Front vulnerable once again.

The Coalition continues to work in support of the Act, in large part debunking myths and misconceptions. Despite eight public meetings, hundreds of personal meetings, and dozens of articles over the past four years, opponents claim the Heritage Act is a government takeover (it's all federal lands already), vanquishes grazing privileges (just the opposite), and was created in secret under the national direction of out-of-touch organizations back East. Some anti-environment attitudes never change regardless of the subject.

But that's our intent regarding the Front. It should never change, at least not at the hands of humans who are looking for a one-night stand with its resources. The dream holds steady: Sustain the Front as it is by living in harmony and respect with its wild inhabitants, holding this place and space for future generations of all species. Dreams can come true.

TAR SANDS, PIPELINES, AND THE THREAT TO FIRST NATIONS

WINONA LADUKE, WITH MARTIN CURRY

THERE ARE FEW routes for tar sands oil to travel from the point of extraction in central Canada to the ports that are gateways to global markets. In 2011 U.S. activists successfully delayed approval of the Keystone XL pipeline, which would have carried tar sands oil to the Gulf of Mexico. But the Northern Gateway pipeline, another proposed project, would travel west from Alberta through British Columbia to Pacific Ocean ports—straight through sensitive watersheds, temperate rainforests, and millennia-old communities of First Nations peoples.

❖

I am writing this story because of a bear—a white bear. The Spirit Bears are white bears in a clan of black bears; one out of every ten of these bears is born pure white. Called *Moksgm'ol* by the Tsimshian people, there are only 400 Spirit Bears in the Great Bear Rainforest in northern British Columbia. Their territory surrounds the town of Kitimat, the proposed end of yet another tar sands pipeline, which means large equipment, pipes, possible spills, and a lot of infrastructure may soon be invading the home of these bears.

In January 2012, just two weeks after President Barack Obama announced that the United States would not move ahead with the proposed Keystone XL pipeline from the Canadian tar sands down to the Gulf Coast, Enbridge Inc. was scrambling to show that its proposed Northern Gateway pipeline was a sure thing—and maybe, just maybe, Enbridge was also hoping to scare U.S. policy makers back to the table for another round of negotiations on the Keystone project.

With some very quick maneuvering after Obama's announcement, the Canadian government set up a National Energy Board (NEB) panel that started hearings on the highly contentious terrain of the proposed Northern Gateway pipeline. It would run from Alberta to

British Columbia, crossing 785 rivers and streams, tunneling through the Coast Range twice, and spanning the headwaters of three of the continent's most important rivers—the Mackenzie, Fraser, and Skeena. The pipe would also punch through the heart of the Spirit Bears' home. Upon reaching the coastal town of Kitimat, the oil would be pumped into holding tanks and then into colossal oil tankers called Very Large Crude Carriers (VLCCs, if you are an insider), which would then chug over the wreckage of a large government-owned passenger ferry—Queen of the North—that is still leaking oil into the ocean, as it has been for six years. After snaking through 120 miles of fjords, making a few tight turns offshore of the largest remaining temperate rainforest on the planet, the VLCCs would arrive at an open ocean that has record-setting tidal fluctuations. Beyond that, it's a clear shot to China. None of that was mentioned at the National Energy Board's Enbridge hearings held thus far.

In a hotel conference room in Edmonton, Alberta, filled with families from Cree communities in northern Alberta, Enbridge officials listened to testimony from members of the communities that would be touched by the Northern Gateway pipeline. Cree villagers talked about their land being overrun with roads and power lines, poisoned by oil and its by-products—and about their rivers and fish already exhibiting signs of stress, and long stretches of days when the fish are inedible and full of tumors. Panel officials politely nodded while Enbridge attorneys and public relations flacks scribbled notes. I sat next to the Enbridge representatives, and as I watched them I recalled what a First Nations chief from British Columbia once told me: "You know what it's like sitting down to negotiate with the Canadian government? It's like sitting down to talk with a cannibal. You can make as much small talk as you want, but in the end, you both know what he is thinking."

The hearings were interesting in a number of ways. Enbridge is pushing for quick regulatory approval of the Northern Gateway pipeline without providing cost and market analysis, adequate assessment of alternatives, or environmental impact studies. Enbridge is pushing for hearings with the support of the NEB, without any discussion of either the broader ramifications of the pipeline, such as the anticipated massive increase in production at the tar sands, or the transport of that oil through innumerable delicate ecosystems. Another intriguing fact is that this would be a pipeline to nowhere, since there are no ready markets for the oil. Despite that, Canadian Prime Minister Steven Harper has said the pipeline is "of national interest."

Peter Okimaw walked up to the microphone. He is a middle-aged man from the village of Driftpile, a Cree reserve around 250 miles north of Edmonton. "When I was a kid, I used to drink from those creeks and rivers," he said. "Now I have to go to Walmart and buy water when I go into the bush by those creeks. When I turned 18, I started to work for the oil industry at Fort McMurray [Alberta]. I cleaned up oil spills. If I tell my son something, it is: Do not work on those oil spills. The last one I worked on was at Slave Lake. We got hired on power-saw operations for ten to twelve hours a day. My brother and I, we were working in power-saw pants. We did not have proper equipment. We were walking in the oil in the creek for ten to twelve hours a day without proper equipment."

During this portion of the testimony, the Enbridge representatives made few notes and wrote very slowly.

Okimaw continued: "I had seen the beaver there going crazy. I said to my brother that we wouldn't go crazy. We would go to our hotel room, and our eyes and our skin would be burning. Now I am wondering if that is going to do something to me in the future. It is all changed now. Like our river, no one swims there now because it is contaminated. Sure we'd like to take our children to swim, but where? We have to go to Edmonton to swim in a pool. It's our traditional land and we should keep it that way. We should save the rest. Once you've taken the heart of Mother Nature, then where do we stand? The world will be looking at us in Alberta. They will say, 'Boy, it was good while it lasted.'"

THE INVISIBLE HAND OF THE INVISIBLE MARKET

Here are a few major problems with the Northern Gateway pipeline project. To begin with it has no customers. It is a pipeline for the sake of building a pipeline. There is a glut in export pipeline capacity in western Canada. Current oil production in western Canada leaves 41 percent of existing export pipelines empty, according to a report by the Natural Resources Defense Council, which says, "Based on industry production estimates, no additional export pipelines are needed out of the tar sands for at least another ten years."

The panicked response in the American press and from some U.S. politicians is that this Canadian oil will be sold to China, but the 525,000 barrels out of the Northern Gateway pipeline would amount to just 6 percent of what China used in 2010, and there is the real possibility that China may be content to buy cheaper oil from Iran. The U.S. Energy Information Administration says, "Iran's best customer is China, which took about 22 percent of Tehran's oil exports during the first half of [2011]" and is "one of the few nations on friendly terms with the Islamic republic. China's demand will continue to grow, but there are a number of sources for this oil."

Is it truly in Canada's "national interest" to spend around $6 billion on a pipeline without committed buyers for that oil? Or is it a Field of Dreams strategy. If you build it, they will come?

SAFETY LAST

Those VLCCs, which can hold 2.2 million barrels of oil and are as large as the Empire State Building, would be deployed to move oil from the pipeline through the port of Kitimat. There could be up to 200 shipments each year, and on every journey the VLCC's captain would have to navigate challenging straits, which have never been traversed by anything that large.

Then there is the pipeline problem. Enbridge's track record for spills is not stellar. According to Enbridge's own data, between 1999 and 2010 there were some 804 reported spills, which released 161,475 barrels of hydrocarbons into wetlands, farms, and waterways —approximately half the oil spilled in the Exxon Valdez disaster.[1] In July 2010, an Enbridge pipeline ruptured near Kalamazoo, Michigan, spilling more than 840,000 gallons of tar sands oil. Enbridge's operators were unaware of the spill at first and did not shut the pipeline down for a full 17 hours.[2] A year after the spill approximately 200 acres of river sediment were still contaminated and a nearly 40-mile stretch of the Kalamazoo River remained closed to the public. Kalamazoo has infrastructure and a relatively large population, so the response there was good by industry standards. It is unlikely that the spill response would be nearly so prompt in the mountain passes or rainforest traversed by the Northern Gateway pipeline.

NIGERIA OF THE NORTH

Prime Minister Harper is from Alberta, where he worked for an oil company. A majority of Alberta's politicians come from oil families, oil money, or families who worked in the oil business. According to Eriel Deranger, a Dene mother and activist from Fort McMurray, in the heart of the tar sands, Alberta is so serious about being a petrol state that the Canadian government has actively sought to limit renewable-energy businesses in the region by changing legislation, making special permits, and limiting access to subsidies.

Here's another way of looking at the Northern Gateway pipeline proposal: Think of Alberta as the Nigeria of the North. (Well, there are a lot more white people in Alberta, and Canada's military hasn't killed anybody to protect the oil business.) Both economies have been increasingly dominated by oil. In 2009 Nigeria exported around 2.1 million barrels of oil per day; Canada exported 1.9 million barrels per day. Environmental regulation of the oil industry in both Nigeria and Alberta is lax, and the industry has been actively opposed by Native people—the Ogoni, in particular, in Nigeria and the Cree in Alberta. In the early 1990s, battles between the Ogoni and the oil companies escalated, with the indigenous people demanding some $10 billion in damages, compensation, and denied royalties. They also called for an "immediate stoppage of environmental degradation." Those demands were answered with military action that led to the deaths of an estimated 2,000 people, including noted Ogoni poet and political leader Ken Saro-Wira.

In Alberta, death and oil have a more subtle relationship. Alberta has the highest suicide rate of any province in Canada—127 percent higher than the rest of Canada, with a 400 percent increase in the past fifteen years. All of this is worsened, arguably, by the tar sands. Consider that an estimated 40 percent of the drivers of equipment are on some sort of illegal drug, which of course means these drugs are now increasingly available in the northern Native communities. According to investigative journalists, approximately $7 million worth of cocaine now travels up Highway 63 every week on transport trucks to the north, which is where the aboriginal people live.

Not surprisingly, since it is a petrol state, Alberta has the largest disparity of wealth of any province in Canada, and the gap there between the average income of a non-Native resident and that of a Native resident is equally appalling—the average Alberta family makes six times the income of an on-reserve Native family. This is somewhat similar to the income disparities in Nigeria. Nigeria is Africa's top oil producer, but the number of Nigerians in dire poverty rose to 61.2 percent in 2010 from 51.6 percent in 2004, the Nigerian National Bureau of Statistics said in a recent report. "It remains a paradox ... that despite the fact that the Nigerian economy is growing, the proportion of Nigerians living in poverty is increasing every year," said Yemi Kale, the head of the Nigerian statistics bureau.

"HARD OF HEARING" HEARING

Back at that hearing held at the hotel in Edmonton: Chief Brenda Sam from Driftpile Cree First Nation steps up to the microphone. She reminds the panel of the history of the province, the treaties, and the duality of responsibility. She also reminds her audience of the long history of problematic relations with settlers, including pervasive sexual abuse in residential schools. "The treaty said we should not molest the newcomers. We kept our end of the bargain, but who molested whom? Think about the residential schools? Who molested whom? Our concerns are about impact. We may have not had the right to live as equals in this unjust society, but we have the right to an opinion."

Next, Gene Chalifoux comes to the podium. She is married to a community member, but, from what I can tell, she came from New Brunswick a very long time ago. She says she has moved in with the Natives, and found it a quite delightful place to live. Her Cree name is Kakiwaksquo, "Dry Meat Woman." She explains that she is proud of learning how to live in the north. She boasts about her trap line of squirrels. "It was just a real exhilaration to be a mean old woman and see all these little squirrels hanging by their neck because I did it myself." And then she laughs. We all laugh. But she also made the crucial point that needs to be pondered throughout Canada. "The people here in front of you signed treaty with the government of Canada about 113 years ago. ... Those people that signed treaty had a belief that they would be treated fairly and that this land would be theirs. ... They got the short end of the stick, the same as my ancestors, the Montauks did. ... If you give these people another eight generations, ... at the rate things are going, they won't exist either. ... If you don't deal with them fairly, they won't exist seven generations from now."

And there it is, a fundamental question for Canada, for all of us: If a people disappear in seven generations does that mean responsibility disappears too?

There is a proposal for a massive pipeline for no apparent reason, and a people may disappear because of it. Make that peoples—the Driftpile Cree First Nation is just one of more than 130 First Nations in western Canada who have publicly stated their opposition to the pipeline and the tankers. At least 70 of them have declared bans on the transport of tar sands crude through their traditional territories. It will be interesting to see if corporations and governments respect these bans.

The Keystone XL pipeline was pitched as a way to provide Americans with oil and much-needed jobs. The Northern Gateway pipeline is being pitched on the promise of sales to foreign markets. Here in Edmonton and in the far north, Alberta's Petrol State is hoping that no one will notice the threat to the aboriginal people, the water, and the Spirit Bear.

SWEET AND SOUR

The Curse of Oil in the Niger Delta

MICHAEL WATTS

SINCE IT BEGAN *producing oil in earnest in 1956, Nigeria has become the poster child for the environmental, social, and economic devastation that can be wrought by unfettered fossil fuel production.*

❖

"Oil is King!" screams the headline of a Nigerian newspaper. And with good reason. Nigeria is the tenth largest producer and the eighth largest exporter of crude oil in the world. Nigerian oil production (crude production and natural gas liquids) is currently running at roughly 2.4 million barrels per day. Roughly two-thirds of production is onshore, the remainder is derived offshore from the continental shelf in both shallow and deep water. The Nigerian government expects proven reserves (in 2010 estimated to be 37.2 billion barrels) to grow to 40 billion by 2020. Nigeria contains the largest natural gas reserves in Africa (likely over 185 trillion cubic feet) and is a global player in the production of liquefied natural gas (LNG).

The enormity of the oil presence in the Niger Delta is hard to fully appreciate. Virtually every inch of the region has been touched by the industry either directly through its operations or indirectly through neglect. Over 6,000 wells have been sunk, roughly one well for every ten-square-kilometer quadrant in the core oil states. There are 606 oil fields (355 on shore) and 1,500 'host communities' with some sort of oil or gas facility or infrastructure. There are 7,000 kilometers of pipelines, 275 flow stations, 10 gas plants, 14 export terminals, 4 refineries, and a massive LNG and gas supply complex. The six-train Bonny LNG facility produces 22 million tons each year. The national oil company, Nigerian National Petroleum Corporation (NNPC), and its joint-venture partners (Shell, Exxon, Mobil, Agip, and Total) directly employ an estimated one hundred thousand people.

Nigeria is an archetypical oil nation. Oil has seeped deeply and indelibly into the political economy of Nigeria. In 2007 over 87 percent of government revenues, 90 percent of foreign exchange earnings, 96 percent of export revenues, and almost half of Gross Domestic Product (GDP) were accounted for by just one commodity: oil. With oil prices hovering around $100 a barrel, oil rents—what economists call "unearned income"—provide the Nigerian exchequer with around $50 billion annually. Nigeria is an oil-state, driven by two cardinal principles: capture oil rents and sow the oil revenues. Like other OPEC countries—by most estimates the 13 OPEC members pocketed over $1 trillion in oil revenues in 2011 alone—Nigeria is currently awash in petrodollars. What this oil wealth has wrought, and is likely to bring, is another question entirely.

An "El Dorado" Nigeria is not. Flying into Port Harcourt or Warri at night—viewing the panorama of harsh gas flares burning bright—conveys a sense of the Dantean universe one is about to enter, the unforgiving, ruthless, and austere world of oil. To compile an inventory of the achievements of Nigerian petro-development is a useful, if dismal, exercise. Eighty-five percent of oil revenues accrue to 1 percent of the population. According to former World Bank President Paul Wolfowitz, at least $100 billion of the $600 billion in oil revenues accrued since 1960 have simply "gone missing." Nigerian anticorruption czar Nuhu Ribadu has claimed that, in 2003, 70 percent of the country's oil wealth was stolen or wasted. Over the period from 1965 to 2004, per capita income fell from $250 to $212 while income distribution deteriorated markedly. Between 1970 and 2000, the number of people subsisting on less than one dollar a day in Nigeria grew from 36 percent to more than 70 percent, from 19 million to a staggering 90 million. Over the last decade GDP per capita and life expectancy have both fallen, according to World Bank estimates. According to the United Nations Development Program (UNDP), Nigeria's rank in terms of the Human Development Index (HDI)—a composite measure of life expectancy, income, and educational attainment—is 158, below Haiti and Congo; over the last thirty years the trend line of the HDI index has been upward, but barely.

Nigeria appears close to the top of virtually everyone's global ranking of corruption, business risk, lack of transparency, fraud, and illicit activity. Nigerian fraud has its own FBI website. Nigeria is not a country, as someone once noted, it is a profession. To suggest, as the International

Monetary Fund has, that $600 billion dollars have contributed to *decline* in the standard of living—that most Nigerians are poorer today than they were in the late colonial period as Nwafejoku Uwadibie says—is mind-boggling, and at the same time it exposes a gigantic failure of leadership and governance. Nigeria has become a model failure. After the discovery of oil in Mongolia, a local leader pronounced: "We do not want to become another Nigeria."

What is on offer in the name of oil development is a catastrophic failure of secular nationalist development. It is sometimes hard to grasp the contours of such a claim. From the vantage point of the Niger Delta—but no less within the barracks of the vast slum worlds of Kano, Port Harcourt, or Lagos—oil development is a pathetic and cruel joke. It's not simply that Nigeria is a sort of Potemkin economy (it is of course); it's the cruel fact that the country has become a perfect storm of waste, corruption, venality, and missed opportunity. To say that Nigeria suffers from corruption—"organized brigandage" is how Ogoni leader Ken Saro-Wiwa (hanged in 1995 for his nonviolent activism) once put it—does not really capture the nature of the beast. Money-laundering and fraud on gargantuan scales, missing billions, and inflated contracts in virtually every aspect of public life, area boys, touts, mobile police all taking their cuts and commissions on the most basic of everyday operations. Perhaps there is no better metaphor for this oil-fueled venality than the stunning fact that huge quantities of oil are simply stolen every day. Over the last five years between 100,000 and 300,000 barrels of oil have been stolen daily (perhaps 10 to 15 percent of national output), organized by a syndicate of 'bunkerers' linking low-level youth operatives and thugs in the creeks to the highest levels of the Nigerian military and political classes and even to people within the oil companies themselves.[1]

Nowhere are the failures more profound and visible than across the oil fields of the Niger Delta. For the vast majority, oil has brought only misery, violence, and a dying ecosystem. A new United Nations report on human development in the delta was unflinching in its assessment: The "appalling development situation" reflects the shameful fact that after a half century of oil development "the vast resources from an international industry have barely touched pervasive local poverty." By conservative oil industry estimates there were almost 7,000 oil spills between 1970 and 2000, more than one each day (the real figure might be twice or three times that number). My back–of–the–envelope calculation suggests that an equivalent of one gallon of oil has been spilled for every hundred square meters of the Niger Delta. Nigeria's gas flaring produced a jaw-dropping 70 million metric tons of carbon emissions per year in 2004, "a substantial proportion of worldwide greenhouse gas" according to the World Bank.[2]

Canalization, dredging, large-scale effluent release, mangrove clearance, massive pollution of surface and groundwater—these are the hallmarks of a half century of oil and gas extraction—all qualified, it needs to be said, by the fact that no serious scientific inventory has ever been conducted because environmental impact assessments by the companies and the government regulatory agencies are practically national secrets. Rates of human-induced erosion at Escravos, Bonny, and Imo can be in excess of 60–70 feet per year. Global climate change and rises in sea level are likely to make the problem much worse. A World Wildlife Fund report released in 2006 simply referred to the Niger Delta as one of the most polluted places on the face of the Earth.

By almost any measure of social achievement, the core producing oil states are a calamity. The United Nations, in the most systematic account of development trends, estimates that between 1996 and 2002 the Human Development Indices actually *fell* in these states. Literacy rates there are barely 40 percent; the proportion of primary school children enrolled is 39 percent, according to a Niger Delta Environmental Survey. The degree of decrepitude in primary schools—whether in Port Harcourt or in the further reaches of the riverine areas—is simply staggering: no desks, no teaching materials, no teachers, and usually no roof. Each secondary school serves 14,700 people in a catchment area of 55 square kilometers.

In Delta State, the doctor-to-population ratio is 1 to 27,000 (and it's 1 to 282,000 in some of the local government areas in Southern Ijaw, Bayelsa State). There is one secondary health care facility for every 131,000 people serving an area of 583 square kilometers. The number of persons per hospital bed is three times higher than the already appalling national average. Electricity is a running joke. Outside the urban areas only 20 percent of settlements are linked to a national grid that does not function in any meaningful sense. A striking discovery of the UN was the fact that local government areas with oil facilities are more likely to have significantly higher indices of human poverty. Were Bayelsa a sovereign state it would, according to its HDI, be ranked roughly 160th in the world.

One of the horrors of the delta is that the ultramodernity of oil sits cheek by jowl with the most unimaginable poverty. Around the massive Escravos oil installation with its barbed wire fences, its security forces, and its comfortable houses are nestled shacks, broken down canoes, and children who will be lucky to reach adulthood. "You will just shake your head," says Ugborodu resident Dorothy Ejuwa, casting an eye on the glare of the nighttime lights of Escravos: "For how long can we remain like this? That is our bitterness."

It is all too easy to be apocalyptic in tone—and to endorse a certain sort of catastrophism that afflicts so much writing about the African continent—but Economic and Financial Crimes Commission (EFCC) Chair Nuhu Ribadu was surely right when he observed in 2006 that the Niger Delta situation was "not being taken seriously" and might "end up like ... Somalia."

The first barrels of Nigerian crude oil destined for the world market departed from Port Harcourt harbor on February 17, 1958. To navigate its way through the shallows of the Bonny River, the 18,000-ton tanker *Hemifusus* left from the Port Harcourt dockside half-full. A shuttle tanker accompanied the *Hemifusus* to Bonny Bar, eight miles from the coast, where another 9,000 tons was pumped into the hold. The oil on board had been discovered in the central Niger Delta in 1956 at Oloibiri, a small, remote creek community near Yenagoa—now the capital of Bayelsa State—located 90 kilometers to the west of Port Harcourt. Wildcatters had begun drilling in 1951 in the northern and eastern reaches of what was then called Eastern Nigeria and finally, on August 3, 1956, discovered oil in commercial quantities in tertiary deposits at 12,000 feet. In its first year of operations Oloibiri produced 5,000 barrels of heavy ('sour') crude oil each day. A year later, the first crude oil pipeline connecting Oloibiri to Kugbo Bay, seven miles distant, came on line. Two hundred-ton barges shuttled the oil to two storage tanks in Port Harcourt; from there the oil was then shipped to the Shellhaven refinery at the mouth of the River Thames. Within a few weeks of its arrival, Nigerian gasoline was fueling automobiles in and around London, the new symbols of postwar British prosperity. The Nigerian oil industry had been born.

When the first helicopters landed in Oloibiri in 1956 to the astonishment of local residents, few could have predicted what was to follow. A camp was quickly built for workers; prefabricated houses, electricity, water, and a new road followed. Shell–BP (as it then was) sunk 17 more wells in Oloibiri and the field came to yield, during its lifetime, over 20 million barrels of crude before oil operations came to a close twenty years after the first discovery. Poverty and capped wellheads are all that remain now.

In the decade that followed, the Nigerian oil industry grew quickly in scale and complexity. A giant field was discovered at Bomu in Ogoniland, west of Port Harcourt in 1958; and Shell–BP, which had acquired 46 oil mining leases covering 15,000 square miles, rapidly expanded its operations across the oil basin. Ten years of feverish activity saw the opening of the Bonny tanker terminal in April 1961, the extension of the pipeline system in 1965 (including the completion of the Trans Niger Pipeline and connecting the oil fields in the western delta near Ughelli to the Bonny export terminal), and the coming on stream of twelve 'giant' oil fields, including the first offshore discovery at Okan near Escravos in 1964. Oil tankers lined the Cawthorne Channel like participants in a local regatta, plying the same waterways that, in the distant past, housed slave ships and palm-oil hulks. By 1967, 300 miles of pipelines had been constructed, and 1.5 million feet of wells had been sunk; output had ballooned to 275,000 barrels per day.

By the first oil boom in 1973, Nigerian oil crude production was comparable to the present day (2.4 million barrels per day), accounting for more than 3.5 percent of world output. Nigeria the oil nation had arrived. Despite the slide into a bloody civil war—the Biafra War from 1967 to 1970—fought on and around its oil fields, the Niger Delta had come of age. Nigeria emerged as a theatre of major significance in the global search for low-cost, high-quality oil. By some industry estimations, the Nigerian exchequer now takes in over $1.5 billion in oil revenues each and every week, supplying a larger share of U.S. crude imports than does Saudi Arabia.

A rusting sign sits next to the 'Christmas tree'—the capped wellhead—at Oloibiri. It is Well No. 1, and the sign reads: "Drilled June 1956. Depth: 12,000 feet (3,700 meters)." It is a monument to an exploit-and-abandon culture, just as Oloibiri itself is a poster child for all of the ills and failed promises of what the great Polish journalist Ryszard Kapuscinski calls "the fairy tale of oil." In the 1960s the town had a population of 10,000; it is now a wretched backwater, a sort of rural slum home to barely 1,000 souls who might as well live in another century. No running water, no electricity, no roads, and no functioning primary school. The creeks have been so heavily dredged, canalized, and polluted that traditional rural livelihoods have been eviscerated. "I have explored for oil in Venezuela and ... Kuwait," said a British engineer, "but I have never seen an oil-rich town as impoverished as Oloibiri." In the last few years the town has been rocked by youth violence; Aso Rock, the armed 'cult group', dethroned the traditional ruler amidst allegations of corruption and half-finished community development projects.

It is a bleak picture, a dark tale of neglect and unremitting misery. Oloibiri, said one local, is now a "useless cast-away snail shell after its meat has been extracted and eaten by the government and SPDC [Shell Petroleum Development Company]." As if to mock the sad fact that Oloibiri is a now a sort of fossil, a piece of detritus cast off by the oil industry, a gaudy plaque dating from a presidential visit in 2001 sits next to Well No. 1. It is a foundation stone for the Oloibiri Oil and Gas Research Institute, and for a museum and library, an homage to Oloibiri and the early history of oil. Noble ideas. But the ground has not been broken, and never will be. Regularly defaced, the plaque is policed by touts looking for a commission from erstwhile visitors who want to record where it all began, the ground zero of Nigeria's oil age.

Commodities define the modern history of the Niger Delta. The delta was the 'Oil Rivers' long before it became a global supplier in the world oil and gas market. Bonny Island near the shores of the Bight of Biafra was a slave port by the seventeenth century and later became a prosperous city-state exporting 25,000 tons of palm oil each year to a surging British industrial economy. One hundred and fifty years later it is home to a massive export terminal and one of the world's largest liquefied natural gas complexes. The great hulks of the Royal Niger Company moored in the estuaries of the Niger Delta in the nineteenth century—serving as consulate, treasury, hospital, prison, and residence—were forerunners of the oil barges, the offshore platforms, and the massive Floating Production Storage and Off-Loading (FPSO) vessels that now populate the delta mangroves and Nigeria's coastal waters.

Mass commodities, with their blood and dirt still attached, have always provided the Niger Delta's point of entry into a world economy, its calling card to the capitalist cosmos. If the commodity is what Karl Marx called the economic cell form of capital, oil is a perfect expression of contemporary capitalism's most basic genetic material. Oil's power as a commodity in the market derives from its two-fold identity. It comes first with its usefulness—its expediency—and then with its price tag. Both seem straightforward and unambiguous, but price—and the pricing of oil—is mysterious, confusing, and bewildering, part of a world of appearances that obscures the operations of the system of which oil is part. Walter Benjamin, the great German critic, said that the commodity has a phantom-like objectivity. Commodities are not what they seem and for this reason are subject to all manner of mystification; they come with their own aura.

Take oil. Kapuscinski was witness to the spectacular oil boom in West Asia during the 1970s. In his book *Shah of Shahs,* he says oil "is a filthy, foul-smelling liquid that squirts obligingly up into the air and falls back to earth as a rustling shower of money." It is a resource that "anesthetizes thought, blurs vision, corrupts. ... [It] kindles extraordinary emotions and hopes, since oil is above all a great temptation." Oil has always been vested with enormous, often magical, powers. It has been called a 'resource curse,' the 'devil's excrement,' the source of the 'Dutch Disease.' Oil distorts the organic, natural course of development. Oil wealth ushers in an economy of hyperconsumption and spectacular excess: bloated shopping malls in Dubai or corrupt Russian 'oilygarchs.' There is even a psychological appellation to describe the condition: Gillette Syndrome. ElDean Kohrs (who coined the term) studied the booming coal town of Gillette, Wyoming, in the 1970s and was witness to how a commodity boom brought a corresponding wave of crime, drugs, violence, and inflation. It would afflict new gas fields of Wyoming, indigenous oil communities in Ecuador, and the rough-and-tumble Russian oil fields of Siberia.

There is inevitably an economic and political form of the Gillette Syndrome. Development guru Jeffrey Sachs believes that every increase in oil dependency (measured as the proportion of exports to Gross National Product) produces a corresponding decrease in economic growth. For Oxford University economist Paul Collier, oil and civil war are evil twins. Oil encourages extortion and looting through "resource predation," and it is the feasibility of predation that determines the economics—that is to say, the funding—of civil war. Others argue that "oil hinders democracy": oil revenues permit low taxes and encourage patronage, thereby dampening pressures for democracy; it endorses despotic rule through bloated militaries; and it creates a class of state-dependents employed in a modern industrial and service sector who are less likely to push for democracy. *New York Times* columnist Thomas Friedman has even identified a First Law of Petropolitics: The higher the average global crude price of oil, the more free speech, free press, fair elections, an independent judiciary, the rule of law, and independent political parties are eroded. Venezuelan President Hugo Chavez is, of course, the law's most devious exponent. In sum, the oil world is cursed—a "carbontocracy" impoverished by its wealth.

There is a ring of truth here. Oil wealth can be, and often is, ill managed. Oil-producing states are among the most corrupt and venal anywhere. The world of oil rents is one of spectacular consumption pushed to its limits. But the language of curse invokes a merciless force for adversity, a sort of commodity determinism vesting oil with capabilities it can neither possess nor dispense. The danger is that the curse substitutes the commodity for the larger truths of capitalism, markets, and politics. Is Nigeria cursed by oil or corruption (or capitalism)? By petroleum or politics (or ethnicity)?

If, as Achille Mbembe says, regions at the epicenter of oil production are regularly torn apart by repeated conflicts, can oil nations escape from under the implacable burden of Black Gold? Must the Niger Delta be yet another casualty?

OUTSOURCING POLLUTION AND ENERGY-INTENSIVE PRODUCTION

VANDANA SHIVA

GLOBALIZATION HAS LARGELY *been seen in the context of the outsourcing of information technologies. But the larger outsourcing that globalization is leading to is the outsourcing of pollution and the energy-intensive production of goods. The corporations and consumers of the rich global North thus bear some responsibility for increased greenhouse gas emissions in the poorer global South.*

❖

GLOBALIZATION, EQUITY, AND CLIMATE CHANGE

Climate change today is global in cause and global in effect. The globalization of economies has outsourced energy-intensive production to countries like China, whose cheap products flood the shelves of supermarkets in the rich North and for the rich in the South. The corporations of the North and the consumers of the North thus have their share of contributions to increased emissions in the countries of the South.

In a globalized economy that runs on fossil fuels, addressing pollution by setting emissions levels for each country is inappropriate for two reasons. First, not all the citizens of a country contribute to the pollution. As a result of becoming the world's factory, China has seen its carbon dioxide (CO_2) emissions rise from 13.8 percent of the world's total in 2005 to 22 percent in 2006; this outstrips the emissions of the United States and puts China in first place worldwide. But while China produced 6.2 billion metric tons of CO_2 in 2006 compared with 5.75 billion metric tons for the United States, its per capita emissions of CO_2 were only 4.7 metric tons , compared with 22 metric tons for the United States.

Further, much of China's 6.2 billion metric tons of CO_2 emissions could just as easily be considered U.S. emissions because U.S. companies have outsourced to China the manufacture of goods that are eventually consumed in the United States. Most of what U.S. superstores sell is produced in China. As Charles Fishman reports,

> Wal-Mart, which in the early 1990s trumpeted its claim to 'Buy America', doubled its imports from China between 1997 and 2002 to $12 billion. In the next two years, Wal-Mart increased Chinese imports again by 50 percent, so that in 2004 the company and its suppliers landed Chinese-made goods in the United States with a wholesale value of $18 billion.[1]

Consumer spending accounts for two-thirds of the U.S. economy—and Walmart dominates retail. During the period between 1997 and 2004, while jobs in U.S. manufacturing fell by 20 percent, Walmart's imports from China increased by 200 percent.[2] Americans are shopping themselves out of jobs while Walmart's everyday lowest price does not include the costs of climate change or the costs of the destruction of China's land, water, and air. As Qin Gang, China's Foreign Ministry spokesperson, said in June 2007, "China is now the factory of the world. Developed countries have transferred a lot of manufacturing to China. What many western consumers wear, live in, even eat, is made in China."[3] Outsourcing of manufacturing is also the outsourcing of pollution. As the outsourcing of manufacturing does not mean that companies outsource profits, the carbon footprint should be counted in the accounts of the company that is profiting, not the country that is bearing the burden of pollution.

This inequity is not limited to U.S. companies. As a recent report by Christian Aid states, "While only 2.13 percent of the world's CO_2 emissions emanate from the U.K.'s domestic economy, through the process of globalization CO_2 is emitted around the world on the U.K.'s behalf in China, India, Africa and elsewhere."[4] While the exact global footprint of U.K. companies is not known, our estimate suggests that emissions associated with the worldwide consumption of the top 100 U.K. company products amounts to 12 to 15 percent of the global total.

In fact, the rural poor in China and India are losing their lands and livelihood for an energy-intensive industrialization. To count them as polluters would be doubly criminal. Moreover, when global corporations outsource to China or India, they need to be responsible for the pollution they carry overseas. Corporations are the more appropriate unit for regulating atmospheric pollution in a globalized economy.

So far, emissions trading schemes have rewarded the polluters by giving them quotas for pollution. And these quotas have allowed them to increase their emissions rather than *decrease* them. What is needed is a carbon tax on corporations—both for their production systems, no matter where their facilities are located, as well as for transport. The global economy as currently organized is destroying local production and promoting long-distance supply on the smallest items of everyday use. Long-distance transport is subsidized while local, low carbon emissions production and distribution is penalized. And the high emissions of long-distance transport have been totally excluded in the Kyoto Protocol. Thus, while the policies of the World Trade Organization (WTO) have led to a huge increase in carbon dioxide emissions due to global transportation, this increase is not even accounted for in emission reduction targets.

We need not concern ourselves with the inevitable corporate cry that regulations destroy markets. After all, even the "market instrument" of carbon trading requires government involvement. The Clean Development Mechanism requires the involvement of the United Nations. Instead of regulating through trade in pollution, governments could regulate through a tax on the use of fossil fuel and through incentives for the use of renewable energy. We can either destroy the conditions for human life on the planet while clinging to "free market" fundamentalism or we can secure our future and create climate justice by requiring that commerce works within the laws of ecological sustainability and within the laws of social justice.

HUNGER FOR ENERGY: THE DIVERSION OF FOOD TO BIOFUEL

Of course, the manufacturing of consumer goods is not the only industrial process that has been aggressively offshored to the detriment of the global poor and the benefit of the global rich.

Biomass has always been used for energy by the global poor, who themselves have largely not participated in the modern fossil fuel economy. With rising oil prices and growing concern about the contribution of fossil fuel use to climate change, there is now a scramble for transforming biomass into liquid fuel for the cars of the global rich. Liquid biofuels are very different from biomass energy for the poor, however. One of the most significant differences is that industrial biofuels divert land and food from the poor to the rich, thus aggravating the hunger crisis.

"First generation" biofuels use soybean and palm oil for biodiesel and corn and sugarcane for ethanol. U.S. and E.U. subsidies and energy policies have encouraged the production of these biofuels, essentially diverting food to fuel, thus contributing to hunger. While industrial biofuels are offered as a renewable energy solution which can address both climate change and peak oil, they require roughly the same amount of energy to produce as they provide. Biofuel production is essentially a net-negative energy system.

In Europe, the E.U. biofuel industry was supported by financial incentives to the tune of 4.4 billion Euros ($6.1 billion). If the target of 10 percent blending of gasoline and diesel with ethanol and biodiesel, respectively, by 2020 has to be met, the industry would be subsidized with 13.7 billion Euros ($18.9 billion) per year. This subsidy pulls food away from the hungry. At least 30 percent of the global food price rise in 2008 was due to biofuels; by 2020 food prices could rise by an additional 76 percent because of such diversion.[5] According to FAO in 2008–2009, 125 million metric tons of cereals were diverted to produce biofuels. Around 40 percent of the corn produced in the United States is being converted into ethanol.[6]

Former World Bank president Paul Wolfowitz said in 2006 that biofuels present "an opportunity to add to the world's supply of energy to meet [an] enormous, growing demand and hopefully to mitigate some of the price effects. It's an opportunity to do so in an environmentally friendly way, in a way that is carbon neutral."[7] In contrast, in 2008 then World Bank president Robert Zoellick stated, "While many worry about filling their gas tanks, many others around the world are struggling to fill their stomachs. And it's getting more and more difficult every day."

Even when food is not used to produce biofuel—or when non-food crops such as jatropha are used—food-growing land is diverted to biofuel production. Our study "Food vs. Fuel"[8] shows how pastures and common lands in Rajasthan, and rice-growing land in the tribal areas of Chhattisgarh, have been appropriated for jatropha cultivation for biodiesel.

There is now discussion about "second generation" biofuels, which use cellulosic biomass rather than food for biofuel. Second generation biofuels use forestry and agricultural by-products, such as wheat straw and corn straw, or crops, such as switchgrass, that can be grown where food crops cannot. However, these biofuels are not expected to fully reach the market before 2018. The challenge is to separate the cellulose from the lignin, and then reduce it to simpler sugars by applying intense heat or strong chemicals.

Second generation biofuels will also have an impact on food. If prices are high enough, there is no guarantee that supposedly nonfood-competitive biofuel crops will not be grown on food-producing land. Moreover, if agricultural by-products like straw are diverted to biofuels, no organic matter will be returned to the soil, thus having a negative impact on soil fertility and food security.

The biofuel market is driven by both the limitless energy appetites of industrial society and the profits that biofuel companies can make. Biofuel markets accounted for $76 billion in sales in 2010, a figure which is expected to rise to $247 billion by 2020 and to $280 billion by 2022.[9] Commodities flow to wherever more profits can be made. They do not respect rights or needs. All they respect is profits. That is why commodification of food has gone hand in hand with its diversion to fuel.

World Cereals Production and "End Use" (million metric tons)[10]

	2007/08	2008/09	% change	Change 2007/08–08/09
Total production	2,132	2,287	+1.3%	+155 mmt
Total utilization*	2,120	2,202	——	——
To food	1,013	1,029	+1.5%	+16 mmt
To animal feed	748	773	+3.3%	+25 mmt
To other uses (including biofuel)	359	401	+11.7%	+42 mmt

*Utilization is a combination of production and the use of stocks from the previous year; stocks of cereals went up from 2007/08 to 2008/09 by about 80 million metric tons.

Thus, despite an increase in food production, there has been food scarcity and food riots in 40 countries. Increasing commodity production does not address hunger; rather, it can aggravate it. In 2008–2009, more cereals produced were used for cattle feed and biofuels (1,107 million tons) than for feeding people (1,013 million metric tons). It is estimated that over the period from 2008 to 2018, biofuel will account for 52 percent of the increased demand for maize and wheat and 32 percent of that for oil seeds.[11]

Industrial biofuels solve neither the energy crisis nor the climate crisis, and they are deepening the food crisis by diverting food to fuel.

Part Six

DEPOWERING
DESTRUCTION

INTRODUCTION

Toward an Energy Economy as if Nature Mattered

It is clear that the problem of generating and deploying energy is central to the conjoined fate of humans and nature. It is equally clear that remaking the current energy economy cannot be accomplished overnight. While the challenges we face are formidable, they are not insoluble with thoughtful action and reorientation of our economic goals based on the understanding that planetary limits are real, and that humanity is part of the community of life (not its raison d'etre).

What, then, is the best "solution" to the energy problem? Ask a roomful of policy wonks that question and the answers will range from "drill, baby, drill" to "more nukes" to "install solar panels here, there, and everywhere." There is certainly no shortage of energy experts advocating this or that program, technology, or energy source. Natural gas boosters promote it as the "bridge" fuel toward renewables. Wind power proponents point to its huge potential. Dam builders tout the ostensibly carbon-free energy to be generated from falling water, and so on.

Most intriguing is when energy experts from across the political spectrum align in their thinking. A case in point is the convergence between military planners working to supply soldiers with energy, security officials focused on reducing vulnerabilities to terrorism, conservationists concerned about the health of the biosphere, and peak oil educators hoping to mitigate the disruptions linked to increasing costs and decreasing availability of energy. For all these constituencies, an identical policy prescription emerges: energy conservation and efficiency efforts combined with locally appropriate distributed generation, especially renewables.

Two of America's leading thinkers on energy, Richard Heinberg and Amory Lovins, whose essays follow, fit easily into that disparate choir, but they present strikingly different ideas about the future. Heinberg anticipates an era of depleting resources dominated by energy scarcity; Lovins outlines an ambitious energy-efficiency agenda oriented toward business profitability and continued economic growth. Interestingly, both viewpoints lead to the conclusion that using less energy more efficiently should be the central organizing principle of a rational energy program.

The subsequent essays—on capping the grid, building an effective political movement to combat climate change, investing in distributed energy generation, halting population growth, and protecting natural areas from the energy machine's onslaught—highlight a range of crucial strategies. They are far from comprehensive, however. There are myriad actions, from personal to international, that can begin to depower the current destructive energy economy and begin remaking it as if nature, people, and the future mattered.

Those policies and behavioral shifts will advance change incrementally. Ratcheting down the system's toxic nature will take many forms—but there is every reason to commence that work as aggressively as possible. The opportunities for increased health, beauty, and wildness are practically limitless, even on a finite planet.

THE CASE FOR CONSERVATION

RICHARD HEINBERG

SOONER OR LATER *we must make conservation the centerpiece of economic and energy policy. Energy conservation is the best strategy for pre-adapting to an energy-constrained future and our best hope for averting economic, social, and ecological ruin. The transition to a more durable and resilient but lower-energy economy will go much better if we plan it, and the shift to a conserver society could hold benefits for people as well as for nature.*

❖

E nergy conservation is our best strategy for pre-adapting to an inevita-bly energy-constrained future. And it may be our only real option for averting economic, social, and ecological ruin. The world will face limits to energy production in the decades ahead regardless of the energy pathway chosen by policy makers. Consider the two extreme options—carbon minimum and carbon maximum.

If we rebuild our global energy infrastructure to minimize carbon emissions, with the aim of combating climate change, this will mean removing incentives and subsidies from oil, coal, and gas and transferring them to renewable energy sources like solar, wind, and geothermal. Where fossil fuels are still used, we will need to capture and bury the carbon dioxide emissions.

We might look to nuclear power for a bit of help along the way, but it likely wouldn't provide much. The Fukushima catastrophe in Japan in 2011 highlighted a host of unresolved safety issues, including spent fuel storage and vulnerability to extended grid power outages. Even ignoring those issues, atomic power is expensive, and supplies of high-grade uranium ore are problematic.

The low-carbon path is littered with other obstacles as well. Solar and wind power are plagued by intermittency, a problem that can be solved only with substantial investment in energy storage or long-distance transmission. Renewables currently account for only a tiny portion of global energy, so the low-carbon path requires a high rate of growth in that expensive sector, and therefore high rates of investment. Governments would have to jump-start the transition with regulations and subsidies—a tough order in a world where most governments are financially overstretched and investment capital is scarce.

For transport, the low-carbon option is even thornier. Biofuels suffer from problems of high cost and the diversion of agricultural land, the transition to electric cars will be expensive and take decades, and electric airliners are not feasible.

Carbon capture and storage will also be costly and will likewise take decades to implement on a meaningful scale. Moreover, the *energy* costs of building and operating an enormous new infrastructure of carbon dioxide pumps, pipelines, and compressors will be substantial, meaning we will be extracting more and more fossil fuels just to produce the same amount of energy useful to society—a big problem if fossil fuels are getting more expensive anyway. So, in the final analysis, a low-carbon future is also very likely to be a lower-energy future.

What if we forget about the climate? This might seem to be the path of least resistance. After all, fossil fuels have a history of being cheap and abundant, and we already have the infrastructure to burn them. If climate mitigation would be expensive and politically contentious, why not just double down on the high-carbon path we're already on, in the pursuit of maximized economic growth? Perhaps, with enough growth, we could afford to overcome whatever problems a changing climate throws in our path.

Not a good option. The quandary we face with a high-carbon energy path can be summed up in the metaphor of the low-hanging fruit. We have extracted the highest quality, cheapest-to-produce, most accessible hydrocarbon resources first, and we have left the lower quality, expensive-to-produce, less accessible resources for later. Well, now it's *later*. Enormous amounts of coal, oil, gas, and other fossil fuels still remain underground, but each new increment will cost significantly more to extract (in terms of both money and energy) than was the case only a decade ago.

After the Deepwater Horizon oil spill of 2010 and the Middle East–North Africa uprisings of 2011, almost no one still believes that oil will he as cheap and plentiful in the future as it was decades ago. For coal, the wake-up call is coming from China—which now burns almost half the world's coal and is starting to import enormous quantities, driving up coal prices worldwide. Meanwhile, recent studies suggest that global coal production will max out in the next few years and start to decline.

New extraction techniques for natural gas (horizontal drilling and "fracking") have temporarily increased supplies of this fuel in the United States, but the companies that specialize in this "unconventional" gas appear to be subsisting on investment capital: Prices are currently too low to enable them to turn much of a profit on production. Costs of production and per-well depletion rates are high, and energy returns on the energy invested in production are low. Recent low prices resulted from a glut of production produced by rampant drilling in 2005–2007, which only made economic sense when gas prices were much higher than they are now. All of this suggests that rosy expectations for what "fracking" can produce over the long term are overblown.

Exotic hydrocarbons like gas hydrates, bitumen ("tar sands"), and kerogen ("oil shale") will require extraordinary effort and investment for their development and will entail environmental risks even higher than those for conventional fossil fuels. That means more expensive energy. Even though the resource base is large, with current technology the nature of these materials means they can be produced only at relatively slow rates.

But if the hydrocarbon molecules are there and society needs the energy, won't we just bite the bullet and come up with whatever levels of investment are required to keep energy flows growing at whatever rate we need them? Not necessarily. As we move toward lower-quality resources (conventional or unconventional), we have to use more energy to acquire energy. As net energy yields decline, both energy and investment capital have to be cannibalized from other sectors of society in order to keep extraction processes expanding. After a certain point, even if gross energy production is still climbing, the amount of energy yielded that is actually useful to society starts to decline anyway. From then on, it will be impossible to increase the amount of economically meaningful energy produced annually no matter what sacrifices we make. And the signs suggest we're not far from that point.

In one sense it matters a great deal whether we choose the low-carbon or the high-carbon path: One way, we lay the groundwork for a sustainable (if modest) energy future; the other, we destabilize Earth's climate, shackle ourselves ever more tightly to energy sources that can only become dirtier and more expensive as time goes on, and condemn myriad other species to extinction.

However, in another sense, it *doesn't* matter which path we choose: With human population numbers growing and energy constraints looming, we will have less energy to burn per capita in the future. Plot any scenario between the low-carbon and high-carbon extremes and that conclusion still holds, which means less energy for transport, for agriculture, and for heating and cooling homes. Less energy for making and using electronic gadgets. Less energy for building and maintaining cities.

Efficiency can help us obtain greater services for each unit of energy expended. Research has been proceeding for decades on how to reduce energy inputs for all sorts of processes and activities. Just one example: The electricity needed for illumination has declined by up to 90 percent due to the introduction first of compact fluorescent light bulbs, and now LED lights. However, efficiency efforts are subject to the law of diminishing returns: We can't make and transport goods with *no* energy, and each step toward greater efficiency typically costs more. Achieving 100 percent efficiency would, in theory, require infinite effort. So while we *can* increase efficiency and reduce total energy consumption, we can't do those things *and* produce continual economic growth at the same time.

Humanity is at a crossroads. Since the Industrial Revolution, cheap and abundant energy has fueled constant economic growth. The only real discussion among the managerial elite was *how to grow the economy*—whether in planned or unplanned ways, whether with sensitivity to the natural world or without.

Now the discussion must center on *how to contract*. So far, that discussion is radioactive—no one wants to touch it. It's hard to imagine a more suicidal strategy for a politician than to base his or her election campaign on the promise of economic contraction. Denial runs deep, but sooner or later reality will expose the delusion that endless growth is possible on a finite planet.

Sooner or later we must make conservation the centerpiece of economic and energy policy. The term "conservation" implies efficiency—

building cars and appliances that use less energy while delivering the same services. But it also means cutting out nonessential uses of energy. Rather than continuing to increase economic demand by stimulating human wants, we must begin to think about how to meet basic human needs with minimum consumption of resources, while discouraging extravagance.

If we move toward renewable and intermittent energy sources, a larger portion of society's effort will have to be spent on processes of energy capture. Energy production will require more land and a greater proportion of society's total labor and investment. We will need more food producers, but fewer managers and salespeople. We will be less mobile, and each of us will own fewer manufactured products—though of higher quality—which we will reuse and repair as long as possible before replacing them.

The transition to a more durable and resilient but lower-energy economy will go much better if we plan it. Wherever it is possible for households and communities to pre-adapt, and wherever clever people are able to show innovative ways of meeting human needs with a minimum of consumption, there will be advantages to be enjoyed and shared.

Much of the current public discussion about our energy future tends to turn on the questions of which alternative energy sources to pursue and how to scale them up. But it is even more important to broadly reconsider how we *use* energy. We must strategize to meet basic human needs while using much less energy in *all* forms. Since this will require major societal effort sustained over decades, it is important to start implementation of conservation strategies well before actual energy shortages appear.

With regard to our food system, it is essential to understand that lower energy inputs will result in the need for increased labor. Thus the energy transition could represent economic opportunity for millions of young farmers. Agricultural production must be adapted to substantially reduced applications of nitrogen fertilizer and chemical pesticides and herbicides since these will grow increasingly expensive as their fossil fuel feedstocks rise in price. And higher transport energy costs mean that food systems must be substantially relocalized.

Transport systems must be adapted to a regime of generally lowered mobility and increased energy efficiency. This would most likely require widespread reliance on walking and bicycling, with remaining motorized transport facilitated by car-share and ride-share programs. Electric vehicles and rail-based public transport systems should be favored, and new highway construction halted.

Reduced overall mobility will require substantial changes in urban design practice and land use policies. Neighborhoods within cities must become more self-contained, and cities must be reintegrated with adjacent productive rural areas. Buildings—including tens of millions of homes in the United States alone—must be retrofitted with insulation to minimize the need for heating and cooling energy. New buildings must require net zero energy input. Incentives for installing residential solar hot water systems, and using solar cookers and clotheslines, should be effective and widespread.

Most new sources of energy will produce electricity—and in the cases of solar and wind, electricity will be produced only intermittently. Electricity storage systems (such as pumped water or compressed air) must be built to overcome at least some of the problems of intermittency. Reconfiguration of electricity grids, distributed generation, and alignment of household and industrial energy usage patterns to fit intermittent power availability are other strategies for adaptation.

The historically close relationship between increasing energy use and economic growth suggests that the global economy probably cannot continue to expand as world energy production falters. Therefore, adaptive measures must include efforts to restructure the economy to meet basic human needs and support improvements in quality of life while reducing debt and reliance on interest and investment income. Family planning must be encouraged, as adding more people to a stagnant or shrinking economy simply means there will be less for everyone.

The costs to ecological integrity and to human health of the ever-increasing scale of society's production and transport systems have become the subject of broadening concern in recent decades. Air and water pollution, resource depletion, soil erosion, and biodiversity loss are just some of those costs. With reduced energy use must come the realization that the scale of our human presence on the planet must be appropriate to the Earth's limited budgets of water, energy, and biological productivity.

Altogether, this will constitute a historic shift away from continual societal growth and toward conservation. It will not be undertaken except by necessity, but necessity is inevitably approaching. Barring some technological miracle, we will have less energy, like it or not. And with less energy, we will no longer be able to operate a consumer society. The kind of society we *will* be able to operate will almost certainly be as different from the industrial society of recent decades as that was from the agrarian society of the nineteenth century.

But suppose this analysis is wrong, or that a new miracle technology appears, and energy proves to be abundant rather than scarce. Even then, conservation makes sense: Increasing energy use leads to greater consumption of natural resources of all kinds, and the degradation of wild natural systems. Sooner or later we must rein in consumption— and since signs of ecological decline are already frighteningly prevalent, sooner is clearly better than later.

The shift to a conserver society could hold benefits for people as well as for nature. As we begin to measure success not by the amount of our consumption, but by the quality of our culture, the beauty of the built environment, and the health of ecosystems, we could end up being significantly happier than we are today, even as we leave a far smaller footprint upon our finite planet. But those benefits will be delayed and diluted for as long as we deny the conservation imperative.

REINVENTING FIRE

AMORY B. LOVINS

FOSSIL FUELS CREATED *modern civilization, but their rising costs—to health, security, and economic progress—are starting to eclipse their benefits, undermining the prosperity and security they enabled. At the same time, technological innovation has quietly been making fossil fuels obsolete. In history's greatest infrastructure shift, spanning the entire economy, humans are inventing a new fire: not dug from below but flowing from above, not scarce but bountiful, and except for a little biofuel, flameless.*

❖

Fire made us human; fossil fuels made us modern. Now we need a new fire that makes us secure, safe, healthy, and durable.

Oil and coal built our civilization. Fossil energy became the foundation of our wealth, the bulwark of our might, the unseen metabolic engine of modern life. Yet this enabler of our civilization, this magic elixir that has enriched and extended the lives of billions, has also begun to make our lives more fearful, insecure, costly, destructive, and dangerous. It puts asthma in our children's lungs and mercury in their lunchbox tuna. Its occasional mishaps can shatter economies. Its wealth and power buy politicians. It drives many of the world's rivalries, corruptions, despotisms, and wars. It is changing the composition of Earth's atmosphere faster than at any time in the past 60 million years.

In short, the rising costs of fossil fuels are starting to eclipse their benefits, undermining the prosperity and security they enabled. Fortunately, these problems are not necessary to endure, either technologically or economically. We can avoid them in ways that tend to *reduce* energy costs—because technological progress has quietly been making fossil fuels obsolete.

What's driving this transformation is basic economics. By 2009, making a dollar of U.S. gross domestic product used 60 percent less oil than in 1975, 63 percent less (directly used) natural gas, 20 percent less electricity, and 50 percent less total energy. Oil is becoming uncompetitive even at low prices before it becomes unavailable even at high prices: Peak oil has emerged in demand before supply. Oil use in the industrialized nations represented by the Organization for Economic Cooperation and Development (OECD) peaked in 2005, U.S. gasoline use in 2007. In 2009, Deutsche Bank said world oil use could peak around 2016.

With today's technologies, it is possible to build uncompromised, safe, roomy, peppy, electric autos. Redesigning the entire U.S. automobile fleet to be superefficient and electrified by 2050 could achieve automotive fuel economy equivalent to 125–250 miles per U.S. gallon (1.0–1.9 L/100 km) and would save oil at an average cost below $18 per saved barrel—just one-fifth of today's world oil price. Buying that efficiency and electrification instead of burning oil to provide the same services from today's and officially forecast autos would save $4 trillion. Such "drilling under Detroit" can win the equivalent of 1.5 Saudi Arabias or half an OPEC, and those "megabarrels" are all domestic, secure, clean, safe, and inexhaustible. The investments required for these four- to eightfold more efficient and oil-free autos and for tripled-efficiency trucks and airplanes could yield a 17 percent internal rate of return (IRR) while greatly reducing risks to the oil and automotive sectors and to the whole economy. The trucks and planes could use advanced biofuels or hydrogen, the trucks could even burn natural gas, but no vehicles would need oil. Despite 90 percent more automobility, 61 percent more flying, and 118 percent more trucking, the most biofuel the United States might need would be less than one-fourth its current use of mobility fuel. That little biofuel could be produced two-thirds from wastes and one-third from cellulosic or algal production that needs no cropland and protects both soil and climate.

Coal—America's #4 source of energy services after efficiency, oil, and gas—is a "dead man walking," says Deutsche Bank's Kevin Parker. Ignoring coal's $180–530 billion annual hidden costs, mainly to public health, America's coal-fired power plants now cost more to run than the cost of displacing them by running existing gas-fired plants more and adopting a level of electrical productivity that ten states, on average, already achieved in 2005. That's why U.S. coal use peaked in 2005, and in 2005–2010, coal lost 25 percent of its share of U.S. electrical services to gas, efficiency, and renewables. Nonnuclear alternatives cheaper than new coal plants could displace U.S. coal power more than 23 times. Once suffices.

Asia is the world leader in adding renewable power; in 2010 only 59 percent of China's net new capacity was coal-fired (versus 38 percent renewable, 2 percent nuclear) and coal's share of China's new generation capacity is shrinking. China's net new orders of coal-fired plants fell by half during 2006–2010. China now leads the world in five renewable energy technologies, and it aims to lead in all of them as the core of its next economy. America, where coal now employs fewer people than wind power, remains politically preoccupied with its previous economy.

Yet solar and wind power have become market winners as their prices plummet—by three-fourths in three years for photovoltaic modules. In roughly 20 states and even in cloudy Holland, entrepreneurs now offer to install solar power on your roof for no money down, and thereby to beat your utility bill. The tipping point where alternatives win on pure price is not decades in the future; it is here and now, forming the fulcrum of economic transformation. Across all energy uses, efficiency and renewables now offer effective, reliable, secure, and affordable replacements for fossil fuel. Rapidly scaling those solutions will define winners and losers between firms—and among nations.

Renewable power, with its lower risk and competitive cost, added half the world's 2008–2010 new generating capacity. In 2010 worldwide, renewable generators other than big hydro dams got $151 billion of private investment and surpassed nuclear power's total installed capacity by adding more than *60 gigawatts*. That much solar capacity is now manufacturable *every year.*

In this global race, the United States' capital, technology, and entrepreneurship equip it for success. Yet it's been held back by lack of coherent vision and overdependence on gridlocked government. In 2010, congressional wrangling helped halve U.S. wind power installations, while China doubled its wind power capacity for the fifth year running and blew past its 2020 wind power target. During 2008–2010, America slipped from #1 to #3 in clean-energy investment, then temporarily rebounded to #1 in 2011 thanks to federal initiatives trying to fill gaps in the wounded capital markets—but those initiatives expire in 2011–2012, while China's policy remains consistent. Since 2005,

U.S. electricity's renewable share crawled from 9 percent to 10 percent while Portugal's soared from 17 percent to 45 percent. Germany, with less sun than Seattle, added more solar power capacity in June 2010 than the United States did in all of 2010, and more in December 2011 than the United States did in all of 2011. By mid-2011, more workers made German solar equipment than made American steel. Germany's efficiency-and-renewables strategy has helped cut its unemployment rate to an eleven-year low.

Japan is moving the same way, but not as fast as India. Brazil and Korea are jumping rapidly into clean energy. As wind power wins power auctions across South America and an unsubsidized solar power plant (the world's most productive) in Chile's northern desert beats the grid price, Chile, with perhaps the world's best portfolio of renewable energy options, is trying to decide whether to let them (and energy efficiency) outcompete traditional projects that are attempting to use political clout to make up for their lack of cost effectiveness.

Worldwide, distributed electricity production is running away with the electric-generation market: About 91 percent of new electricity in 2008 came from renewables (excluding big hydro dams) and combined-heat-and-power. All renewables now deliver a fifth of the world's electricity from a fourth of the world's generating capacity. In 2011, the clean-energy market won $260 billion of investment and attracted its trillionth dollar since 2004. All countries hoping to build or retain economic dynamism must catch up with this multi-trillion-dollar, once-in-a-civilization business opportunity.

How? Use our most effective institutions—private enterprise, coevolving with civil society and sped by military innovation—to end-run ineffective institutions like the U.S. Congress. In autumn 2011, such a strategy was detailed in Rocky Mountain Institute's independent energy vision for American leadership, *Reinventing Fire*, with forewords by the CEO of Shell Oil and the chairman of Exelon. Its fresh competitive strategies can win the clean energy race, not forced by public policy but led by business for durable advantage.

Reinventing Fire maps market based paths for running a 158 percent–bigger U.S. economy in the year 2050 (an assumed growth target based on official projections, not personal preferences) with no oil, no coal, no nuclear energy, one-third less natural gas, and no new inventions.

Moreover, this could be accomplished at a net-present-value cost of $5 trillion *below* business as usual, assuming all externalities are valued at zero (a conservatively low estimate, as oil's hidden economic and military costs alone exceed $1.5 trillion a year, excluding any damage to public health and environment).

The business case for efficiency is so compelling that adopting it would require no new federal taxes, subsidies, mandates, *or laws*. Policy innovations that unlock and speed the transition could be implemented with no Act of Congress—instead by federal administrative actions and at the state level, where utilities are already largely regulated (but 36 states still reward them for selling more energy and penalize them for cutting your bill). The key automotive reform could also be readily adopted by states: *Feebates*, a revenue-neutral way to help auto buyers use societal discount rates, tripled the speed of improving new French autos' efficiency in just two years.

If General Dwight Eisenhower couldn't solve a problem, he made it bigger, expanding its boundaries until added options and synergies made it soluble. In the same vein, *Reinventing Fire* integrates all four energy-using sectors—transportation, buildings, industry, and electricity—and four kinds of innovations—technology, design, policy, and business strategy. Together these are much more than the sum of the parts.

The auto and electricity problems, for example, are easier to solve together than separately. New design and manufacturing methods can make ultralight, ultrasafe autos cost-competitive. Needing half or a third the power for the same pep then lets electric propulsion compete too. (BMW, VW, and Audi plan to mass-produce electrified carbon-fiber cars by 2013.) But carbon fiber and electrification are cheaper when combined: three steep and synergistic learning curves—in carbon fiber, automaking, and electric power trains—together create a game changer as potent as the shift from typewriters to computers.

Adding tripled-efficiency trucks and planes, and using all vehicles more productively, enables greatly expanded mobility fueled by a mixture of electricity, hydrogen, and advanced biofuels, but needing no oil. Smart vehicles, buildings, and grids could make electric autos not a burden but a valuable flexibility and storage resource. That is, by buying electricity from the grid or selling it back at the right times, a smart electric auto fleet can help smooth out variations in solar and wind power generation, reducing the need for fossil-fueled generation and making an 80–100 percent renewables-powered electricity grid reliable and competitive.

Doubled energy productivity in industry (with a 21 percent IRR), tripled or quadrupled in buildings (33 percent IRR), can profitably shrink electricity demand despite 84 percent more industrial production and 70 percent more floor space. Just investing $0.5 trillion to fix buildings, which use three-fourths of U.S. electricity, can save $1.9 trillion. The recent retrofitting of the Empire State Building and its resultant two-fifths energy savings with a three-year payback illustrate how integrative design can often yield *expanding* returns, making big energy savings cheaper than small ones: Remanufacturing all 6,514 windows onsite in a temporary window factory on a vacant floor made them pass light but block heat, reducing winter heat loss by two-thirds and summer heat gain by half. Adding better daylighting, lighting systems and controls, and office equipment saved a third of air conditioning on hot days. This in turn saved $17 million of capital cost because the old chillers could be renewed and reduced rather than replaced and enlarged. That capital saving helped pay for the other improvements, cutting the payback to three years. Applying this approach to a twenty-year-old glass office tower could even save three-fourths of its energy, slightly cheaper than the normal twenty-year renovation that saves almost nothing.

The key to this economic magic is "integrative design"—designing a building, factory, device, or vehicle as a whole system and optimizing it for multiple benefits, rather than optimizing isolated components for single benefits. For example, the middle of my own house, high in the Rockies where outdoor temperatures used to fall as low as −47°F (−44°C), is currently ripening its 37th through 39th banana crops with no furnace during a January snowstorm. The house is about 99 percent passive-solar heated, but the superwindows, superinsulation, and ventilation heat recovery that eliminated its heating system added less construction cost than eliminating the heating system saved. Respending that saved capital cost plus a bit more also saved about 90 percent of the household electricity and 99 percent of the water-heating energy, all with a ten-month payback using 1983 technology. Today's technology is much better, so we've just retrofitted it and are measuring its performance; unfortunately, the monitoring system seems to be using more electricity than the lights and appliances it's measuring.

An even more striking example comes from pumping—the main use of motors, which use three-fifths of the world's electricity. Using fat, short, straight pipes rather than narrow, long, crooked pipes saves typically 80–90 percent of the friction in the pipes. Shrinking the pumps, motors, inverters, and electrical systems more than pays for the fatter pipes, decreasing total capital cost. In my own house, this tactic cut friction by about 97 percent. Fans and ducts, the second biggest use of motors, offer similar opportunities. And every unit of friction saved in pipes or ducts saves about ten times more fuel, cost, and what Hunter Lovins calls "global weirding" back at the power station.

Industry is already ripe in opportunities for better motor systems and pumps, fans and controls, heat recovery and insulation. Dow Chemical has already saved $19 billion on $1 billion of efficiency investments. But integrative design can make savings bigger yet cheaper, turning diminishing returns into expanding returns. Rocky Mountain Institute's latest $30-odd billion worth of integrative redesign of equipment and processes across diverse industries—from refineries to mines and data centers to chip fabs—has typically reduced expected energy use by about 30–60 percent with a few years' payback on retrofits, or by about 40–90+ percent with generally lower capital cost in new factories. Integrative design isn't yet included in official studies of energy-saving potential, but smart firms are realizing how it can drive competitive advantage. RMI's Factor Ten Engineering (10xE) initiative aims to use it to transform how design is done and taught.

Combining modern ways to wring more work out of each kilowatt-hour could power a 2.6-fold bigger U.S. economy with one-fourth less electricity than now, eliminating not just coal-fired but also nuclear power production. That's good, because as those old plants retire (virtually all by 2050), replacing them with more of the same would be so costly and risky that no business case can be made for it. All 34 new proposed nuclear plants in the United States can't raise any private capital despite 100+ percent construction subsidies: At most a few units may be built, entirely financed by mandatory payments from customers and taxpayers. All 66 reactors under construction worldwide at the end of 2010 were bought by centrally planned power systems. Nuclear power's death of an incurable attack of market forces strengthens climate protection, because new nuclear plants are so costly and slow that they would save about 10–20 times less carbon per dollar, about 20–40 times slower, than investing in efficiency and renewables instead.

Productive and timely use of electricity, combined-heat-and-power, reallocated saved natural gas, and a modern renewables portfolio can enable a diverse, distributed, reliable, resilient electricity future that costs about the same as business as usual but manages all its risks, including economy-shattering blackouts. Replacing America's aging, dirty, obsolescent, insecure electricity system by 2050 will cost about $6 trillion in net present value no matter how we do it. So let's re-architect and rebuild it to power not just lights and motors but also competitive advantage, profits, jobs, national security, environmental stewardship, and public health, while making the grid so resilient that big cascading blackouts become impossible. Whichever of those outcomes you care most about, *Reinventing Fire*'s pragmatic business strategy makes sense and makes money.

Recent experience and practice also confirm that even with little or no bulk power storage, diversified and forecastable renewable generators, integrated with flexible supply and voluntarily modulated demand, can deliver highly reliable power at competitive cost. Four German states in 2010 got 43–52 percent of their electricity from wind power by integrating it with the strong German grid. But even on a continental scale, diverse renewables can provide 80+ percent of electricity by operating utilities' existing assets differently within smarter grids and using markets that clear faster and serve larger areas.

Reinventing Fire's U.S. findings are highly adaptable and adoptable elsewhere. The European Climate Foundation has presented a similarly ambitious road map for Europe's energy transition, as have many countries. Governments from California (the world's #8 economy) to Germany (#4) and from Denmark to Sweden are successfully implementing aggressive efficiency-and-renewables strategies. California shrank greenhouse gas emissions per dollar of GDP by 30 percent in 1990 2006, and has held per capita use of electricity flat for three decades while real income per capita grew by four-fifths. Denmark's GDP grew by two thirds during 1980–2009 while energy use fell back to its 1980 level and carbon emissions fell 21 percent. In an average wind year, Denmark in 2010 could produce 36 percent of its electricity renewably and 53 percent from combined-heat-and-power. The average Dane, releasing half the carbon of the average American, enjoyed a good life, the most reliable electricity in Europe, and some of its lowest pretax prices. Denmark is even reorganizing its grid in "cellular" fashion (as Cuba successfully did) to make power supply highly resilient—and plans to be entirely off fossil fuels by 2050.

Developing countries are buying the majority of the world's new renewable generating capacity, often in distributed forms like solar cells that bring efficient lighting and other vital services to the 1.6 billion humans who have no electricity—leapfrogging over the power line phase just as cell phones leapt past landline phones. If developing countries buy efficiency whenever it's cheaper than new electricity supply, they can turn the power sector, which now devours a fourth of global development capital, into a net exporter of capital to fund other development needs. Why? Because making super-efficient lamps, windows, and the like takes about a thousand times less capital, and repays it about ten times faster, than investing instead in supplying more electricity. Investing in cheap "negawatts" instead of costly megawatts is the most powerful, though invisible, financial lever available to speed global development.

The international Super-efficient Equipment and Appliance Deployment project (SEAD), supported by 23 countries, targets the four appliances—lights, refrigerators, air conditioners, and televisions—that use three-fifths of household electricity in China, India, the United States, and the European Union. Most of those appliances haven't yet been built or bought, and three-fourths are made by just 15 firms. SEAD aims to build them right, saving up to $1 trillion and avoiding 300 coal plants. That's just four household appliances—not the rest, not the other sectors. There are lots more negawatts to capture.

China made energy efficiency its top strategic priority in 2005, not compelled by a treaty but because leaders like Wen Jiabao understood that China couldn't afford to develop otherwise. China had already fueled about 70 percent of 1980–2001 economic growth by cutting energy intensity more than 5 percent per year, and is now regaining that pace. The United States has long averaged 2–4 percent lower energy intensity (primary energy used per dollar of GDP) each year. Just averaging 3–4 percent worldwide could prevent further climate damage. Why should that pace be hard, since most of the growth is in countries like China and India that are building their infrastructure from scratch and can more easily build it right than fix it later? And since virtually everyone who does energy efficiency makes money, why should this be costly?

The global climate debate has focused on cost, burden, and sacrifice because negotiators assumed from economic theory that energy efficiency must cost more than the energy it saves, or we'd have bought it already in their theoretically perfect markets. But actually, most efficiency isn't yet bought, even in the most competitive market economies, because of 60–80 kinds of widespread and well-documented market failures that we now know how to turn into business opportunities. In truth, *saving fuel costs less than buying fuel, so climate protection is not costly but profitable.* Talking instead about the resulting profits, jobs, and competitive advantages so sweetens the conversation that any remaining resistance should melt faster than the glaciers.

In history's greatest infrastructure shift, humans are verily inventing a new fire: not dug from below but flowing from above, not scarce but bountiful, not local but ubiquitous, not transient but permanent, not costly but free—and except for a little biofuel, grown in ways that sustain and endure, flameless.

Efficiently used, this new fire, harnessed by ingenuity and enterprise, can make energy do our work without working our undoing. The new energy era can be a story not of danger, restriction, and impoverishment but of astounding wealth creation, choice, and opportunity.

CAP THE GRID

ROBERT E. KING

AS A SPECIES, *we must learn to live within the physical limitations of the biosphere. In the electric energy sector, this requires reversing the worldwide trend of ever-expanding electricity supply grids carrying energy vast distances from more and more large, centralized power plants. "Capping the grid" is a crucial step toward reducing greenhouse gas pollution and increasing the percentage of electricity generated by renewables.*

❖

The electric grid, comprised of generation, transmission, and distribution facilities, is controlled in the United States by "independent system operators" and electric utilities. These grid operators work closely with power plant owners and government regulators as they operate and plan for the future of the electric system. Grid operators regularly forecast electricity usage in their region and then set off alarm bells when they see the forecasted demand for electricity exceeding the supply. There follows a Pavlovian reaction as planners, engineers, and generation companies enter a bidding process to determine who will build the next round of generation and transmission to meet the predicted supply shortfall.

Civic leaders respond to the dire scenarios forecast and become part of a positive feedback loop driving an ever-expanding grid by pushing for additional power plant and transmission construction. The end result, not surprisingly, is to keep fossil fuel use high, electricity supplies plentiful, prices relatively low, and serious conservation limited only to the virtuous.

Grid growth results from population growth, economic growth, and growth in per capita consumption. In a finite global ecosystem, societies would do well to figure out how to flourish in the absence of unchecked growth on all these fronts. They *must*, however, figure out

how to stop—and ideally reverse—grid growth because of the multi-
tude of negative impacts associated with the expanding grid. The U.S.
Energy Information Administration predicts the grid will grow in the
next twenty-five years at an annual rate of 0.5–1 percent. This may not
sound like much, but it would require adding ten power plants the size
of Three Mile Island *every year.*

Specific drivers of grid growth include the Internet, which now consumes 5–10 percent of worldwide electrical energy. Countless servers and
personal computers running 24/7 have a larger carbon footprint than the
world's aviation industry, according to recent research. In the near future,
electric cars, which are more efficient than gas or fuel cell vehicles, will
demand an increasing share of the grid's output, though they can also help
stabilize the grid if programmed to send energy back from their batteries
at peak demand hours. If car charging occurs late at night, it takes advantage of existing generating capacity when it is underutilized. To make
this a truly ecological progression, however, the charging energy should
be derived from deep conservation, such as the retirement of millions of
unnecessary lights that now currently pollute the night sky in the interest
of 24-hour commerce and presumed security.

Since the 1978 passage of the Public Utility Regulatory Policies Act,
with many fits and starts, renewable energy sources have begun to fill in
the need for additional energy supply. Their obvious advantages include
lower carbon and other emissions and reduced dependence on foreign oil
and gas. But every renewable resource comes with an environmental cost.
Hydropower of all sizes affects river hydrology and biota. Large hydro
typically destroys vast and increasingly rare wild places. Industrial-scale
wind can similarly erode wilderness values while harming avifauna and
possibly humans. Biomass quickly runs into the ecological limits of forests that are often better left standing. Concentrated solar power plants
can destroy desert habitat, and manufacturing solar photovoltaic panels is
energy intensive and produces toxic pollution.

If we simply add renewables to an ever growing energy mix, then we
have the negative impacts of renewables in addition to the carbon emissions of the fossil-based system. Clearly we have to rein in this system-
wide growth, and two essential elements for doing so are a cap on energy
generation and a price for the external costs of this generation (primarily
carbon emissions). A cap would insure that each and every megawatt-hour

of generation by renewables permanently displaces the same amount of fossil-fueled generation. For example, if the controversial Cape Wind offshore wind farm in Massachusetts is brought online, the dirty and inefficient fossil-fueled Mirant Canal Power Plant, also located on Cape Cod, should be retired, thereby permanently reducing the carbon emissions of the regional grid.

In addition, the external costs of electric generation must be internalized. Putting a price on carbon emissions creates a further incentive to retire dirty fossil generators like Mirant Canal. Shutting down old plants as renewables ramp up will also help prevent a supply glut that would put downward pressure on the price of electricity. Energy planners and economists will find this idea heretical because they believe that increasing prices will be deleterious to the economy. What really matters to the frugal consumer, however, is not the cost per kilowatt-hour but the bottom line on his or her electric bill. The individual or business consumer will pursue deep energy conservation not because someone is telling them to, but because it's the obvious solution when faced with rising energy prices—prices that reflect the true cost of the energy.

Implementing aggressive conservation and energy efficiency across all energy sectors should be the first order of business in national energy policy, even before development and adoption of large-scale renewables. People often think that energy conservation means replacing incandescent bulbs with compact fluorescents. This is the tip of the melting iceberg. Efficiency is changing a light bulb. Conservation is never installing the light fixture in the first place. Efficiency is when you buy a TV with an LED screen. Conservation is when you jettison the TV. But efficiency and conservation by themselves are not enough.

If efficiency measures are implemented by a large cohort of early adopters, in the absence of other market or regulatory signals the outcome might fall short. The reason is that a significant reduction in usage across the grid will lower demand and result in a throttling back of the more expensive power plants that operate on the supply margin. The spot market price of electricity will drop, and the average consumer will have incentive to use more, not less energy (an example of what is known as Jevons Paradox). A 2010 study by Sandia National Labs reached a related conclusion—that super-efficient LED lighting may in fact result in more light usage and possibly more electricity usage overall. Throughout history, people have endeavored to add more artificial light to their daily lives. With compact fluorescents and LEDs, people think their light source is more efficient, so they may be prone to leave it on longer or add more lighting. The solution is to slowly and steadily increase per unit energy prices in parallel with increasing efficiency. The wise consumer will attempt to reduce his usage ahead of increasing energy prices.

Energy use per capita and energy intensity (energy input per unit of economic output) have started to decline in some states and countries, particularly when price signals are combined with conservation programs. A wise national energy policy would embrace this decline and improve upon it dramatically to achieve an energy sector that does not place an undue burden on the climate and all natural systems. A good example is California, where energy efficiency has been the mandated energy resource of first choice since 2003. Utility companies' profits are decoupled from energy sales, and now they earn additional profits by implementing energy efficiency programs. Over the last twenty years, increasing efficiency has replaced power plant construction even as the population and the economy have grown. One reason that efficiency measures have worked is that prices were allowed to creep up. So California has reversed the trend of ever-increasing energy intensity. It is building renewables aggressively and often with significant environmental impact, but the state's carbon footprint within the electricity sector is shrinking.

The least destructive energy future lies in conservation, efficiency, centralized and dispersed renewables, and an ongoing use of natural gas to transition through the carbon and population bottleneck of the next fifty years. Photovoltaics should be installed on every rooftop in this country within twenty-five years, paid for in part by a permanent and hefty tax credit and further rationalized by generally higher energy prices. Large-scale wind in the plains, mountains, and offshore will be critical, although the preservation of wild and beautiful places must take precedence. Biomass-fueled electricity will be less plentiful, assuming we don't wish to overcut forests and convert too much land to agriculture for biofuel crops, but will be useful in some regions as baseload generation. Natural gas, though a carbon emitter, will remain in the mix as baseload power since it has the potential to be more efficient and less polluting than other fossil fuels. (To achieve its promise as a "bridge fuel" to a clean energy future, however, gas extraction through hydrofracturing must be effectively regulated and the leakage of unburned methane system-wide must be eliminated.) All these supplies will be augmented by centralized and distributed energy storage facilities to smooth the volatile curve of small-scale renewable energy production.

How is all this to come about? It will only occur via accurate price signals and strong regulation. It will come when an engaged citizenry says "Enough!" to endless government subsidies of entrenched technologies: enough of loan guarantees and the Price Anderson Act, limiting liability at nuclear plants; enough subsidies to Big Oil via Middle East military engagements; enough subsidies to all fossil fuel use by allowing the dumping of carbon and other pollutants into the atmosphere for free. Then, upon a level playing field, renewables will easily compete.

Perhaps this economic internalization of externalities presently borne by society and nature will come as a tax shift toward carbon and away from income, as has begun in some European countries. Perhaps it will be the carbon "Fee and Dividend" approach advocated by NASA scientist James Hansen, in which a price on carbon is collected and refunded to taxpayers. Germany's rooftop solar subsidies known as Feed-In Tariffs have dramatically grown its solar industry and renewable generating capacity. The United States flirts with incentives, but for these to be truly effective they should be made permanent so that the renewables industry is not subject to constant uncertainty. Tax credits are a step in the right direction, but a complete overhaul of the tax system to internalize carbon emissions and other externalities and plow some of the revenues back into efficiency has the potential to radically alter our energy future.

The current energy economy cannot be fundamentally reformed until society discards the idea that endless growth in the energy sector is a prerequisite for "the good life." We should not begin to develop the large-scale renewables envisioned above *unless every new, renewable megawatt-hour will replace a dirty, carbon-emitting megawatt-hour.* Imposing a moratorium on overall growth of the electricity generation sector is crucial. Then, through scarcity, prices will increase and simple laws of supply and demand will usher in the largest conservation and efficiency movement ever seen. An endlessly expanding energy sector, fueled by nuclear and other mega-industrial developments, only perpetuates the impossible premise that there are no limits to how much energy we should consume.

As a practical matter, how do we move "cap the grid" from idea to reality? Grid operators control the electric system, but they are subject to oversight by the Federal Energy Regulatory Commission and various reliability councils. Ultimately, all these organizations respond to political pressure, which could be applied by a revitalized energy reform movement. In the Northeast, the Regional Greenhouse Gas Initiative, a cap and trade program for carbon emissions, is now the law because the governors and legislators of ten states got together and agreed to make it so. It took time, and it is not nearly as strong as it needs to be, but change in the electric sector is possible if enough people demand it.

How would a grid cap work? A cap would be applied to the total energy production on the grid. With an increasing price on carbon, the oldest, dirtiest plants would become uneconomic, making way for cleaner generation. In their planning processes, grid operators and regulators would allow for new generation as the older, dirtier plants were retired. Of course, the embedded technocrats who have built today's grid will put up tremendous resistance. Overcoming this institutional momentum will require a unified environmental movement, which now wastes resources on internecine battles over the impacts of various renewable energy projects. Perhaps there would be more tolerance for some of these projects if it was recognized that *every megawatt-hour of renewables would permanently replace a megawatt-hour of carbon-based generation.*

A grid cap does not mean that no more transmission lines will ever be built. New or higher voltage lines could be built in existing transmission corridors, and local distribution systems could be improved to accommodate distributed generation and storage. Europe has been much more serious about its carbon reduction strategies; the European Commission's recently developed "Energy Roadmap 2050" envisions the necessary infrastructure for a renewably powered Europe. It is instructive because it massively scales up renewables. It is formidable in its call for thousands of miles of underwater high-voltage direct-current transmission lines as a way to link the wind and hydro resources of Scandanavia with the solar potential of the Mediterranean countries. There are similar calls in the United States for a huge increase in terrestrial transmission capacity to carry wind energy from the Great Plains to the coasts.

It is not clear if the environmental impacts of large, new transmission projects are an acceptable trade-off for the benefits of dispersed renewables. And there are other problems with the use of long transmission lines. In the United States, about 6.5 percent of the energy flowing into the grid is lost—turned into heat in the wires. This equates to the energy used by 24 million average American households! By contrast, when energy is generated on your rooftop by solar panels, the transmission loss is essentially zero.

New transmission lines through undeveloped landscapes should be avoided entirely. These are typically vast linear clear-cuts with subsequent reductions in carbon sequestration, and they often act as gateways to industrialism in our wildest and most remote places. Whether on land or undersea, thousands of miles of high-voltage lines through remote areas represent a large national security risk. They are unprotected and more vulnerable than most policy makers realize. As recent blackouts have shown, minor failures in key nodes on the grid can take out power to entire regions. Transmission line construction is enormously expensive—another cost ultimately picked up by the electricity consumer. Thus, new transmission projects are and will be hugely controversial, and there is no point in entertaining them if we cannot first agree to "cap the grid."

What about the Smart Grid? Will it solve our energy problems? The Smart Grid is happening already and it does have some potential benefits, particularly if politicians will allow price signals to play a role. One benefit will come when commercial and industrial consumers are subject to real-time energy pricing and can easily access this information along with their usage data via smart meters. In this case, they would quickly figure out when to scale back usage, thereby correcting grid overloads before they happen. Instead, regulators now mandate extra generation and transmission capacity to meet the highest peaks, even though this capacity is often idle outside of peak periods. But the wonders of the Smart Grid are easily oversold. The Energy Independence and Security Act of 2007 compels some grid operators to begin using the satellite GPS network to collect grid data that informs grid control. Is this more complicated system really more *independent or secure*? No, it is simply an escalating dependence on multiple, interwoven layers of vulnerable technology.

Real security and independence would come from deep conservation and energy efficiency and decentralized renewable generation, with less dependence on massive, long-distance energy flows. Lasting security and independence would come from a system of pricing that reflects all the costs of our energy choices and thereby motivates consumers such that serious conservation would no longer be the realm of the virtuous.

PROTECTED AREAS

Foundation of a Better Future Relationship with Energy

HARVEY LOCKE

PROTECTING AREAS FROM *resource extraction is the one sure way to address the paradox that energy production and consumption are both powering and destroying our civilization. At least half of planet Earth's land and surface waters should be in protected areas dedicated to nature conservation to conserve biological diversity and stabilize the climate. Securing "Nature's half" of the world in interconnected protected areas would help humanity develop a sane relationship with energy.*

❖

Vast energy consumption powers modern civilization and has brought humans many benefits; it is also serving to destroy wilderness, a stable climate, and the conditions for human well-being. Thus we have created a paradox: Energy production and consumption are at once powering and destroying our civilization. There is one sure method to shift this dynamic that is good for both nature and human well-being—protecting areas from resource extraction. And I mean lots of protected areas: protected areas that are interconnected; protected areas on land and sea; protected areas that conserve at least half of the land and water on planet Earth. We should do this because human life depends on it, because the rest of life depends on it, and because it is the right thing to do.

We and the rest of life are made up of carbon. Carbon comes in several forms: living (you and me, trees, sea organisms), ancient (coal, oil, natural gas, limestone), and gas (carbon dioxide, methane). There is a dynamic exchange between these forms of carbon. Trees pull carbon dioxide out of the atmosphere. We breathe it out into the atmosphere. Some of it settles into the oceans to become limestone or is taken up by living things like phytoplankton, kelp, and mangroves. When brackish seashore environments get buried, their carbon-rich vegetation is eventually compressed and turns into coal. Organisms become oil and gas through a similar process. This transformation from living carbon to carbon stored in rocks takes a very long time.

Carbon in various forms is the dominant source of energy for humans. Humans have long burned living carbon from trees in campfires and thus converted it to gas. We also burned oil from living things such as whales for lighting, heating, and cooking. In the eighteenth century we significantly increased wood burning and began burning coal for a new purpose, industrial development. But it wasn't until the last half of the nineteenth century that we learned also how to burn oil from rocks. (It was originally called "rock oil" and was once drunk for its alleged health benefits; "petroleum" refers to rocks.) This incredibly efficient energy source that we now know as oil laid the foundation for the industrialization of the world and modern transportation systems. These activities also radically altered the dynamic exchange among forms of carbon, which is called the "carbon cycle."

In the last two hundred fifty years we have removed and burned so many forests, we have converted so many grasslands and wetlands, and we have disturbed so many mangroves that we have converted an enormous amount of living carbon into atmospheric gas (carbon dioxide). One-quarter to one-third of the great rise in carbon dioxide in the atmosphere since 1750 that is driving climate change has come from converting living carbon to gas. Most of the other two-thirds of the great rise in carbon dioxide since the Industrial Revolution has come from burning ancient carbon for energy and a small portion that comes from burning limestone to make cement. Today the ratio is more like 20 percent of all carbon dioxide coming from converting living systems and 80 percent from burning ancient carbon. All of this is radically transforming the climate in dangerous ways.

The net effect of this conversion of living and ancient carbon to atmospheric gas was neatly summarized by the Intergovernmental Panel on Climate Change in 2007: "The resilience of many ecosystems is likely to be exceeded this century by an unprecedented combination of climate change, associated disturbances (e.g., flooding, drought, wildfire, insects, ocean acidification), and other global change drivers (e.g., land use change, pollution, overexploitation of resources)." In other words, business as usual is a disaster for nature, which is also a disaster for us.

Humans depend on nature for life. We do not know how to make water or air. It comes from nature. All the food we eat comes from nature. While we know how to manipulate life to increase yields, we simply do not know how to make life. Ecosystems and their interaction with the atmosphere, freshwater and salt water, and the minerals from the Earth are the foundation of human life. We definitely do not want to overwhelm them. Yet that is exactly what we are doing.

Young people know this is happening. The generations of humans that have been in positions of power and leadership during the past thirty years have done something that is very intergenerationally unfair. We have compromised our children's future. We have deprived them of hope for a better life. It doesn't matter that we did not mean to do it. We did it. And we continue to do it despite the overwhelming evidence that it is self-destructive. We don't seem able to stop ourselves. Young people will have to live with the consequences.

There is an obvious solution that can materially address for the better every aspect of this nasty problem that we have created: protected areas, where industrial-scale extraction of living or ancient carbon and other minerals is not permitted. Examples are national parks, wildlife reserves, watershed protection areas, municipal nature parks, tribal sacred areas, and wilderness areas. They can be owned by the public, private individuals, or associations, or be tribally controlled.

Protected areas work because they stop land use change, they stop overharvesting, and they do not contribute pollution. They prevent greenhouse gas emissions in two ways: They end the destruction of natural ecosystems immediately, which prevents greenhouse gas emissions from living carbon; and they keep the Earth's ancient carbon in place, which keeps it out of the atmosphere. Protected areas also provide for resilience through allowing ecosystems to adapt to climate change, especially if they are large and interconnected across elevations and latitudes. They are the single best tool we have to conserve ecosystems and reduce carbon emissions. And they are beautiful.

Canada, the United States, Australia, Argentina, countries of the European Union, South and East Africa, and many others have long histories of setting aside exceptionally beautiful places as protected areas. Protected areas have worked well, but we do not have enough of them and they alone are not sufficient to secure the future of life. They must protect the full variety of habitats (not only the exceptionally scenic) and be connected across the landscape to allow species to move and find mates and to allow plants and animals to adapt to climate change.

To date we have protected less than 15 percent of the planet's land. Even less of the sea is protected. While countries like Australia and the United Sates have made some important recent advances in creating marine protected areas, many more are urgently needed to curtail the continuing widespread decimation of fish populations and destruction of reef structures by industrial fleets and conversion of coastal mangroves to shrimp farms. By committing at least half the world's lands and waters to protection of nature and ensuring they are interconnected, we can largely overcome the problems of overexploitation, fragmentation, and isolation of plant and animal species, while effectively keeping huge amounts of living and ancient carbon out of the atmosphere.

Of course civilization today depends on access to and use of ancient carbon. So we cannot stop burning it immediately; a transition will be necessary. We could, however, move to protect at least half of the Earth now, where carbon extraction is not under way. Antarctica is an area of unclear or shared sovereignty; all of it should be protected forever. The Arctic Ocean is an area of disputed sovereignty; let's solve the problem by agreeing to protect it. The same applies to the high seas where through international actions we could protect at least half right away. Countries with vast de facto wilderness like Canada, Australia, and Russia could act immediately to protect at least half of their landscapes and seascapes. Similarly we should protect at least half (and probably much more) of Patagonia and all of the remaining tropical rainforests on Earth.

Why protect at least half the world? The science is clear that what is necessary to allow for ecosystems to retain all their species, continue the natural processes they support, and be able to adapt to environmental change is to protect at least half of a given ecological system in an interconnected manner. So let's do what is needed.

Creating protected areas at this scale is already under way. Australia recently announced that it will protect nearly half of its southeastern coast. Ontario and Quebec, the two largest provinces in Canada, have formal plans to protect half of their northern regions (in Quebec alone this means that 135 million acres, an area roughly the size of France, will be protected from all industrial development). In the North American west, the Yellowstone to Yukon corridor is showing that protecting and interconnecting a vast area can be done while people prosper. Suriname has protected a vast area of tropical rainforest.

In western Europe and eastern North America there has actually been a recovery of forests that were previously cleared. Following the model of Adirondack State Park in New York, we should protect them now and allow them to grow "forever wild" to fix more carbon and house more life. In places that have been more radically transformed by humans, we should vigorously pursue ecosystem restoration and then protect the areas we restore. Coastal mangroves and sea grass beds—among the most biologically productive ecosystems and largest carbon storehouses on Earth—are a particularly good place to start. This will take considerable time, as mangroves, once cleared, are slow to recover.

People like me who propose simple solutions to intertwined problems are often attacked for failing to understand "reality." This is a shame. Sometimes the answer to complex problems is really simple: Stop focusing on all the facets of complexity and instead address the bigger underlying problems. Protected areas work because leaving nature intact is the single most important thing we can do. We know how to establish protected areas and how to manage them. We can create alternate economic opportunities for local people who are affected by them. Further, protected areas don't cost much to create. They will force us to do more with less. And they will help solve the problems in our relationship with energy that we can't seem to solve otherwise. We just need the will to act.

We should act immediately to protect at least half the world in an interconnected manner. This would show we are serious about making the twenty-first century a good one for life. It would help us develop a sane relationship with energy. It would provide an agenda of hope for future generations. To secure at least half of the world in interconnected protected areas with sustainable development on the rest would be a project worthy of a great civilization.

THREE STEPS TO ESTABLISH A POLITICS OF GLOBAL WARMING

BILL MCKIBBEN

DESPITE INCREASINGLY WORRYING *scientific evidence, worsening extreme weather disasters, and years of advocacy by the major environmental groups, political leaders in the United States have not acted seriously on climate change. Because efforts to push climate change action through regular political channels have clearly failed, a mass movement of grassroots citizen activism is necessary.*

❖

TRY TO FIT THESE FACTS TOGETHER:

- In early 2009, southeastern Australia experienced the most extreme heat wave in its recorded history; later that year, a dust storm over 2,000 miles long blanketed most of the country's east coast, choking major cities in a red haze.
- In 2010, a "staggering" new study from Canadian researchers showed that warmer seawater has reduced phytoplankton, the base of the marine food chain, by 40 percent since 1950.[1] Also in 2010, carbon dioxide emissions made the biggest one-year jump ever recorded (5.9 percent) and pushed atmospheric carbon dioxide concentrations to 389.6 parts per million, the highest level in the last eight hundred thousand years.
- In 2011, the United States saw 14 "billion-dollar" extreme weather events, more than ever before, causing more than 600 deaths and over $52 billion in damage. Among the worst were a devastating drought and wildfires in Texas, flooding from Hurricane Irene in the Northeast, major flooding in the Midwest and along the Mississippi River, and six different multiday tornado outbreaks in the Midwest and Southeast.

And during all this time, our leaders did exactly nothing about climate change. The 2009 U.N. Climate Change Conference in Copenhagen turned out to be an elaborate sham: We discovered afterwards that negotiators knew their proposed emissions cuts were nowhere near good enough to meet their own remarkably weak target for limiting global warming. In 2010, Democratic Senate majority leader Harry Reid decided not even to schedule a vote on legislation that would have capped carbon emissions. And in 2011, the presidential campaign started off with the major Republican candidates falling over each other to proclaim their disbelief in human-caused climate change.

I wrote the first book for a general audience on global warming back in 1989, and I've spent more than two decades working on the issue. I'm a mild-mannered guy, a Methodist Sunday School teacher. Not quick to anger. So what I want to say is this: *The time has come to get mad, and then to get busy.*

For many years, the lobbying fight for climate legislation on Capitol Hill has been led by a collection of the most corporate and moderate environmental groups. We owe them a great debt, and not just for their hard work. We owe them a debt because they did everything the way you're supposed to: they wore nice clothes, lobbied tirelessly, and compromised at every turn.

By the time they were done, they had a bill that only capped carbon emissions from electric utilities (not factories or cars) and was so laden with gifts for industry that if you listened closely you could actually hear the oinking. They bent over backwards like Soviet gymnasts. Senator John Kerry, the legislator they worked most closely with, issued this rallying cry as the final negotiations began: "We believe we have compromised significantly, and we're prepared to compromise further."

And even that was not enough. They were left out to dry by everyone—not just Reid, not just the Republicans. Even President Obama wouldn't lend a hand, investing not a penny of his political capital in the fight.

The result: total defeat, no moral victories.

So now we know what we didn't before: Making nice doesn't work. It was worth a try, and I'm completely serious when I say I'm grateful they made the effort, but it didn't even come close to working. So we better try something else.

Step one involves actually talking about global warming. For years now, the accepted wisdom in the best green circles was: talk about anything else—energy independence, oil security, beating the Chinese to renewable technology. I was at a session convened by the White House early in the Obama administration where some polling guru solemnly explained that "green jobs" polled better than "cutting carbon."

No, really? In the end, though, all these focus-group favorites are secondary. The task at hand is keeping the planet from melting. We need everyone—beginning with the president—to start explaining that basic fact at every turn.

It *is* the heat, and also the humidity. Since warm air holds more water than cold, the atmosphere is about 5 percent moister than it was forty years ago, which explains the freak downpours that seem to happen someplace on this continent every few days.

It *is* the carbon—that's why the seas are turning acid, a point Obama could have made with ease while standing on the shores of the Gulf of Mexico. "It's bad that it's black out there," he might have said, "but even if that oil had made it safely ashore and been burned in our cars, it would still be wrecking the oceans." Energy independence is nice, but you need a livable planet to be energy independent on.

Mysteriously enough, this seems to be a particularly hard point for smart people to grasp. Even in the wake of the disastrous Senate non-vote, a climate expert from one of the big green groups told *New York Times* columnist Tom Friedman, "We have to take climate change out of the atmosphere, bring it down to earth, and show how it matters in people's everyday lives."[2] Translation: ordinary average people can't possibly recognize the real stakes here, so let's put it in language they can understand, which is about their most immediate interests. It's both untrue, as I'll show below, and incredibly patronizing. It is, however, exactly what we've been doing for a decade and clearly, It Does Not Work.

Step two, we have to ask for what we actually need, not what we calculate we might possibly be able to get. If we're going to slow global warming in the very short time available to us, then we don't actually need an incredibly complicated legislative scheme that gives door prizes to every interested industry and turns the whole operation over to Goldman Sachs to run.[3] We need a stiff price on carbon, set by the scientific understanding that we can't still be burning black rocks a couple of decades hence. That undoubtedly means upending the future business plans of ExxonMobil and BP, Peabody Coal and Duke Energy, not to speak of everyone else who's made a fortune by treating the atmosphere as an open sewer for the by-products of their main business.

Instead they should pay through the nose for that sewer, and here's the crucial thing: *most of the money raised in the process should be returned directly to American pockets*. The monthly check sent to Americans would help fortify us against the rise in energy costs, and we'd still be getting the price signal at the pump to stop driving that SUV and start insulating the house. We also need to make real federal investments in energy research and development, to help drive down the price of alternatives—the Breakthrough Institute points out, quite rightly, that we're crazy to spend more of our tax dollars on research into new drone aircraft and Mars orbiters than we do on solar photovoltaics.[4]

Yes, these things are politically hard, but they're not impossible. A politician who really cared could certainly use, say, the platform offered by the White House to sell a plan that taxed BP and actually gave the money to ordinary Americans.

Asking for what you need doesn't mean you'll get all of it. Compromise still happens. But as David Brower, the greatest environmentalist of the late twentieth century, explained amid the fight to save the Grand Canyon: "We are to hold fast to what we believe is right, fight for it, and find allies and adduce all possible arguments for our cause. If we cannot find enough vigor in us or them to win, then let someone else propose the compromise. We thereupon work hard to coax it our way. We become a nucleus around which the strongest force can build and function."[5]

Which leads to the third step in this process. If we're going to get any of this done, we're going to need a movement, the one thing we haven't had. For twenty years environmentalists have operated on the notion that we'd get action if we simply had scientists explain to politicians and CEOs that our current ways were ending the Holocene, the current geological epoch.[6] That turns out, quite conclusively, not to work. We need to be able to explain that their current ways will end something they actually care about, i.e., their careers. And since we'll never have the cash to compete with ExxonMobil, we better work in the currencies we can muster: bodies, spirit, passion.

As Tom Friedman put it in a strong column the day after the Senate punt, the problem was that the public "never got mobilized."[7] Is it possible to get people out in the streets demanding action about climate change?

In 2009, with almost no money, our scruffy little outfit, 350.org, managed to organize what *Foreign Policy* called the "largest ever coordinated global rally of any kind" on any issue—5,200 demonstrations in 181 countries, 2,000 of them in the U.S.A.[8]

People were rallying not just about climate change, but around a remarkably wonky scientific data point, 350 parts per million carbon dioxide, which NASA's James Hansen and his colleagues have demonstrated is the most we can have in the atmosphere if we want a planet "similar to the one on which civilization developed and to which life on earth is adapted."[9] Which, come to think of it, we do. And the "we," in this case, was not rich white folks. If you look at the 25,000 pictures in our Flickr account,[10] you'll see that most of these citizens were poor, black, brown, Asian, and young—because that's what most of the world is. No need for vice presidents of big conservation groups to patronize them: shrimpers in Louisiana and women in burqas and priests in Orthodox churches and slum dwellers in Mombasa turned out to be completely capable of understanding the threat to the future.

Those demonstrations were just a start (one we should have made long ago). We followed up in October 2010—on 10-10-10—with a Global Work Party. All around the country and the world people put up solar panels and dug community gardens and laid out bike paths. Not because we can stop climate change one bike path at a time, but because we need to make a sharp political point to our leaders: We're getting to work, what about you? We need to shame them, starting now. And we need everyone working together.

This movement is starting to emerge on many fronts, beginning with grassroots citizen opposition to mountaintop-removal coal mining, new coal plants, the shale gas drilling boom, and the unlocking of Canada's tar sands. In 2008, a young man named Tim DeChristopher tried to keep treasured public lands out of the hands of energy developers (and that underground carbon out of the atmosphere) at a rushed BLM gas-and-oil lease auction in the waning days of the Bush Administration. Tim posed as a bidder, won 14 parcels, and got ten years in prison for his efforts—but barely two months later new Interior Secretary Ken Salazar canceled most of the auction's sales.

In 2011, tens of thousands of people across the continent rallied to stop Keystone XL, a massive new pipeline meant to bring nearly 1 million barrels of tar sands oil daily from Canada to refineries and ports in the United States. We marched, lobbied, even circled the White House with our bodies, and over 1,000 of us were arrested—but President Obama ultimately sent the project back for more review (a modest but real victory).

The big environmental groups are starting to wake up, too. They did amazing work on the Keystone campaign, rallying people for unconventional action and working easily and powerfully with the grass roots—groups like the Natural Resources Defense Council, Friends of the Earth, the Sierra Club, and the National Wildlife Federation were out in front, aggressive, impassioned. Churches are getting involved, as well as mosques and synagogues. Kids are leading the fight,[11] all over the world—they have to live on this planet for another seventy years or so, and they have every right to be pissed off.

But, it won't work overnight. We're not going to get Congress to act next week, or maybe even next year. It took a decade after the Montgomery bus boycott to get the Voting Rights Act. But if there hadn't been a movement, then the Voting Rights Act would have passed in…never. We may need to get arrested again. We definitely need art, and music, and disciplined, nonviolent, but very real anger.

Mostly, we need to tell the truth, resolutely and constantly. Fossil fuel is wrecking the one Earth we've got. It's not going to go away because we ask politely. If we want a world that works, we're going to have to raise our voices.

DISTRIBUTED RENEWABLE GENERATION

Why It Should Be the Centerpiece of U.S. Energy Policy

SHEILA BOWERS AND BILL POWERS

INDUSTRIAL-SCALE WIND *and solar power projects can produce significant quantities of renewable energy, but distributed renewable energy generation— particularly rooftop photovoltaic installations—can achieve the same objective much faster without the environmental harm and at lower cost. With state and federal policies that favor distributed energy, the United States could greatly expand the direct involvement of individuals and communities in renewable power generation.*

❖

Distributed electricity generation (local, decentralized energy production) has the potential to radically alter America's energy land-scape. Our current energy mix is dominated by large, remote, centralized power facilities such as nuclear, gas, and coal-fired power plants, as well as massive wind farms and transmission infrastructure. Today, improved technologies, environmental and economic concerns, and a recognition of the vulnerabilities in large centralized power production make distributed generation coupled with efficiency upgrades a viable and, in fact, preferable alternative. Every properly situated building, parking lot, and brownfield (disused, contaminated land) in our communities can potentially become a producer of energy.

Distributed generation most commonly involves solar photovoltaic (PV), but can also include small hydroelectric, small-scale biomass facilities, and micro-wind. There are several advantages to distributed generation when good policies are implemented. Foremost is that the bulk of the economic benefits of widely distributed, locally controlled, and locally produced clean energy can go directly to ratepayer-generators and property owners through mechanisms such as the feed-in tariff, a generous per-kilowatt-hour payment made to ratepayers who generate clean power on their homes and businesses. Additionally, distributed energy generators often enjoy substantial improvements in property values, according to the Appraisal Institute.

Remote, centralized power production and its associated transmission are substantially more vulnerable to major electrical shutdowns from earthquakes, hurricanes, fires, wind, ice, human error, cyber attack, or terrorism than distributed generation (which connects to the local power grid). Because of local redundancies and geographic diversity, a well-designed local grid with distributed power production and adequate storage can reliably provide critical energy in times of storms or emergencies with less disruption and pollution than conventional solutions. Perhaps most importantly, millions of acres of healthy, intact ecosystems are left undisturbed when generation is sited within the built environment.

With the proper incentives and policies (such as a German style feed-in tariff) distributed energy can be built much more quickly than large centralized power facilities and their attendant transmission infrastructure. For instance, German residents installed 7,400 megawatts (MW) of local, rooftop solar PV in 2010 alone, at an installed cost substantially lower than the projected installed cost of utility-scale solar thermal or PV power plants sited in U.S. deserts or arid grasslands (not one of which came online in 2010 or 2011). And the pace of such installation is accelerating; in just one month (December 2011) 3,000 MW of solar PV was added to Germany's energy portfolio.

Distributed energy production also makes multiple uses of urban and suburban landscapes, including rooftops, and can provide incentives to remediate brownfields that would otherwise blight neighborhoods for decades. Solar photovoltaic sited within the built environment, as well as properly sited micro-wind power, can be deployed without disrupting natural communities; without government using its powers of eminent domain; without depleting scarce groundwater; without destroying viewsheds and recreation areas; and without the waste and destruction that has become the hallmark of the energy industry.

Because distributed energy can be locally produced, locally owned, and locally consumed—bringing both economic benefits and jobs to communities—there is typically less local opposition to implementing distributed energy projects than to building (and financing) centralized, large-scale power projects. Large-scale renewable energy projects—such as most proposed solar power plants and industrial wind generation sited in remote locations—represent a continuation of the old paradigm of large-scale industrial development, owned and controlled by monopoly interests which externalize the majority of their costs onto ratepayers, taxpayers, and the environment while privatizing the profits. In contrast, small-scale projects are often strongly supported by local communities. San Francisco recently pledged to procure 100 percent of its electricity from local renewable energy, as has a growing collection of European towns.

Finally, distributed energy is already technologically feasible, even more so when coupled with efficiency upgrades and passive heating/cooling systems for buildings. Energy consulting giant KEMA recently reported that the California grid is capable of very substantial increases in local solar generation without expensive grid upgrades.

DISTRIBUTED SOLAR POWER: RUNNING THE NUMBERS

When all costs are factored in—including new transmission infrastructure and line/heat losses—local, distributed solar PV is comparable in efficiency, faster to bring online, less destructive, and less expensive than remote utility-scale solar plants.

Net energy output of rooftop solar is comparable to utility-scale desert solar.

"Higher solar insolation," meaning higher solar radiant energy, is the most common reason touted for siting utility-scale solar projects in locations like the Mojave Desert. However, transmission losses largely negate the benefits of such remote projects compared to the slightly lower solar insolation of ratepayer regions like Los Angeles, Riverside, and San Diego (which are required to purchase the solar power). Power transmission losses average 7.5–14 percent in California[1], but the difference in solar insolation between the Mojave and Southern California urban centers is approximately 10 percent.[2] This means there is no substantial difference in the net electric power delivered to customers from remote utility-scale solar plants in remote Mojave Desert locations and rooftop PV installations in Riverside or Los Angeles, for projects with the same rated capacity.[3] Urban rooftop solar has another distinct

advantage: Desert solar production drops precipitously at higher temperatures, when power is needed most, because both PV and air-cooling are
less effective at high temperatures.

Rooftop solar is faster to implement.

Large-scale remote solar projects and related transmission lines take many
years to permit and complete. In contrast, distributed PV can be brought
online very quickly. Germany, using a simple and effective feed-in tariff contract structure to spur cost-effective development of distributed
PV, installed 7,400 MW of distributed PV in 2010 alone, 80 percent of
it locally owned and sited within the built environment—a 75 percent
increase from 2009. These results are consistent with previous years, and
compare very favorably to the less than 900 MW of PV installed in the
United States in 2010, a country with a population roughly three and a
half times as large.

Rooftop solar is a more economically sound investment.

California's Renewable Energy Transmission Initiative reported that, in
comparing May 2010 prices for solar thermal and PV, the latter had a
cheaper cost per megawatt-hour of electricity production. Solar photovoltaic prices have dropped substantially since then, while solar thermal
costs have risen or remained static. This reality is reflected in the shift, en
masse, by utility-scale solar developers away from thermal projects and
toward PV, the exact technology that is used for distributed generation.

In late 2011 the residential rooftop solar consolidator 1 Block Off
the Grid reported actual installed costs (prior to any rebates, tax credits,
or other incentives) at $4.18 per watt in New Jersey. Installed PV system
costs in other areas of the country, from Massachusetts to California,
are less than $5 per watt, and both the California Energy Commission
and the Department of Energy project that solar PV prices will drop by
half between 2010 and 2020, while solar thermal prices are projected to
decline much more gradually, if at all.[4] A study done by the Los Angeles
Business Council and the University of California–Los Angeles estimated that there may be enough rooftops in Los Angeles County suitable for solar to produce roughly 19,000 MW. It also found that there is
at least 3,300 MW of rooftop solar currently "economically available"
for German-style feed-in tariffs for the City of Los Angeles alone, and

it estimated that their proposed 600 MW feed-in tariff program would create more than 11,000 local jobs.[5] The feed-in tariffs required to provide a fair return on investment for ratepayer-generators would cost ratepayers very little. The study projected an average monthly additional cost of only $0.48 per month for households and $9.37 per month for businesses for the first ten years of the program, after which point ratepayers would enjoy *lower* electricity bills than if they had remained with conventional energy.

New transmission infrastructure needed to carry utility-scale solar-generated energy from remote locations to urban demand centers also entails substantial costs that distributed generation does not. These costs are ultimately borne by ratepayers, with actual costs for new California transmission lines currently running from approximately $11 million to $24 million per mile.[6] In addition, rooftop solar creates local, well-paid, long-term jobs; substantially improves property values; encourages energy conservation; and, when supported by common-sense mechanisms such as feed-in tariffs, slows the outflow of cash to utilities, keeping money in communities.

Industrial solar enjoys enormous subsidies and externalized costs.

Large-scale remote solar projects enjoy a number of direct and indirect subsidies that are not available to the ratepayer-generator, putting the latter at an enormous disadvantage. These often include federal cash grants and very low-interest loans and loan guarantees; exclusive use of public lands, water, and resources otherwise designated for multiple uses; waivers of millions of dollars in application fees; extremely high Power Purchase Agreement (PPA) prices; and externalization of many types of costs onto local communities, ratepayers, and ecosystems.

Rooftop solar reduces greenhouse gas emissions faster and more effectively.

Unlike energy systems in the "concrete jungle" ecosystem, large, remote solar projects permanently reduce natural uptake of carbon by the ecosystems cleared for development, while also releasing the carbon dioxide that had been sequestered by them. Researchers at the University of Nevada–Las Vegas have been monitoring carbon uptake in Mojave Desert ecosystems for several years and have consistently found substantial sequestration of carbon.[7] Likewise, wetland and grassland ecosystems such as those found in Colorado's San Luis Valley (targeted for industrial solar development) are well-known for their ability to uptake and store carbon dioxide.[8] More study is needed to determine how much carbon uptake will be lost when hundreds of thousands of acres of natural desert cover are converted to scraped earth and covered with solar collectors, but it is safe to assume that it is more than "none" which is the case with distributed generation.

JUMP-STARTING SOLAR PV

Distributed generation, supported through feed-in tariffs, Property Assessed Clean Energy (PACE) loan financing, and greatly expanded net metering would be more effective than remote utility-scale solar in producing reliable, affordable, nondestructive renewable energy and addressing the climate crisis. Feed-in tariffs are proven to work quickly, economically, and reliably; they provide a simple contract mechanism for individual homeowners and business owners to profitably install as much solar PV as their buildings/properties will allow, maximizing the potential of rooftops, parking areas, and brownfields in urban and suburban environments. Even as its solar PV tariffs shrink, Germany continues to increase the amount of PV installed—largely because of the rapid decline in the cost of PV systems, which is built into the design of the feed-in tariffs. Thanks to these effective cost-reduction policies, it is currently less expensive, on a per-watt-installed basis, to install a custom, small, rooftop solar system in Germany than it is to install a giant, ground-mounted desert solar installation in the United States, despite economies of scale.

PACE loans, which allow ratepayers to amortize the costs of rooftop solar and efficiency upgrades over many years and repay them along with property tax payments, have proven very popular where offered. There is no cost to taxpayers or other ratepayers, and virtually no risk to lenders or borrowers. Unfortunately, the Federal Housing Finance Agency has suggested that because property tax assessments take the first lien on an applicable property, PACE loans will pose a threat to the supremacy of mortgages held by their agencies.[9] This has effectively crushed residential PACE lending. Restoring this critical tool would boost democratically owned clean energy considerably.

CONCLUSION

A growing body of research and analysis favorably demonstrates the economic, environmental, and community benefits of local distributed energy compared to remote utility-scale energy of all types. With policies that favor distributed clean energy generation and efficiency, the United States could avoid the elevated economic and ecological costs of remote utility-scale power development, preserve critical wildlife habitat, reduce greenhouse gas emissions, create jobs, and gain significant economic benefits for local residents, businesses, and communities.

NO ECOLOGICAL SUSTAINABILITY
WITHOUT LIMITS TO GROWTH

PHILIP CAFARO

THE MAIN CAUSE *of global climate change is growth: unrelenting economic growth and a swelling human population. Most proposals for reducing climate change take this growth for granted and focus exclusively on technical means for reducing greenhouse gas emissions. An economic system dedicated to generating ever more wealth and consumption for ever more people must be replaced by an economy designed to provide sufficient wealth and consumption for a finite number of people.*

❖

There is a curious disconnect in climate change discussions between explanations of the causes of global climate change and discussion of possible solutions. On the one hand, it is widely acknowledged that the primary causes of climate change are unrelenting economic growth and swelling populations. As the fourth assessment report from the Intergovernmental Panel on Climate Change (IPCC) succinctly puts it: "GDP [Gross Domestic Product] per capita and population growth were the main drivers of the increase in global emissions during the last three decades of the twentieth century.... At the global scale, declining carbon and energy intensities have been unable to offset income effects and population growth and, consequently, carbon emissions have risen."[1] On the other hand, most proposals for reducing climate change take this growth for granted and focus exclusively on technical means for reducing greenhouse gas emissions.

Climate scientists speak of the "Kaya Identity," which describes the relation of the four primary factors that determine anthropogenic greenhouse gas emissions. These are economic growth per capita, population, energy used to generate each unit of GDP, and greenhouse gases generated per unit of energy. Over the past three and a half decades, improvements in energy and carbon efficiency have been overwhelmed by increases in population and wealth. Here are the numbers, again according to the IPCC:

> The global average growth rate of carbon dioxide emissions between 1970 and 2004 of 1.9 percent per year is the result of the following annual growth rates:
> - Population, +1.6 percent,
> - GDP per capita, +1.8 percent,
> - Energy intensity (total primary energy supply [TPES] per unit of GDP), −1.2 percent,
> - And carbon intensity (carbon dioxide emissions per unit of TPES), −0.2 percent.[2]

Importantly, the IPCC's projections for the next several decades see a continuation of these trends. More people living more affluently mean that under "business as usual," despite expected technical efficiency improvements, greenhouse gas emissions will increase between 25 percent and 90 percent by 2030, relative to 2000.[3] If we allow this to occur, it will almost surely lock in global temperature increases of more than 2°C over pre-industrial levels, exceeding the threshold beyond which scientists speak of potentially catastrophic climate change. I believe following this path would represent a moral catastrophe as well: the selfish over-appropriation and degradation of key environmental services by the current generation to the detriment of future ones, by rich people to the detriment of the poor, and by human beings to the great detriment of the rest of the living world.[4]

A reasonable person reading the IPCC reports and related scientific literature on climate change would likely conclude that humanity is bumping up against physical and ecological limits. Facing catastrophic global climate change, a prudent and moral response might be: *Wow! This is going to be hard. We need to start working on this problem with all the tools at our disposal. Increasing energy and carbon efficiency, to be sure. Let's get the techno-wizards working overtime on those. But also decreasing the pursuit of affluence and overall consumption, and stabilizing or reducing human populations. Maybe in the future we can grow like gangbusters again, although that's doubtful (and really, why would we want to?). But for now, people need to make fewer demands on nature and see if even our current numbers are sustainable over the long haul. After all, our situation is unprecedented—7 billion people living or aspiring to live in modern, industrialized economies—and we may already be in 'overshoot' mode.*

Such convictions would only be strengthened by considering further evidence of global ecological degradation from the 2005 *Millennium Ecosystem Assessment* (MEA), including the depletion of ocean fisheries, accelerating soil erosion, ongoing species extinctions throughout the world, the growth of immense "dead zones" at the mouths of many great rivers, and more. According to the MEA, humanity is currently degrading or utilizing unsustainably 15 of 24 key ecosystem services.[5] For the morally obtuse, a visit to Appalachia's ravaged mountains and streams, or Alberta's degraded tar sands mining areas, or Louisiana's oil-soaked marshes might provide further incentive for change.

However, neither global warming nor the worldwide damage of fossil fuel production has led to a widespread reevaluation of the goodness of growth.[6] Regarding climate change, we have seen a near-total focus on technological solutions by politicians and even by scientists and environmentalists, who should know better.

Numerous illustrations can be cited from the IPCC's fourth assessment report itself. Its authors recognize agriculture as a major contributor to global climate change, for example. Yet they simply accept projections for greatly increased demands for all categories of agricultural products (including a doubling in worldwide demand for meat over the next fifty years) and focus on changes in tillage, fertilizer use, and the like as means to limit increased greenhouse gas emissions.[7] Similarly, the assessment report notes that among significant greenhouse gas sources, aviation traffic is the fastest-growing sector worldwide. It considers numerous changes to aviation practices, including relatively trivial improvements in airplane technology and changes in how high planes might fly on particular routes, while avoiding the obvious alternative of reducing the number of flights.[8] Many similar examples could be given.

All this is a serious mistake. Because "business as usual" with respect to growth undermines our attempts to avoid catastrophic global climate change or meet our other global ecological challenges, we almost certainly need to slow or end growth. "Can advances in science and technology prevent global warming?" asks Pacific Northwest National Laboratory scientist Michael Huesemann in a recent review article of the same name. After detailed analysis, he answers that an exclusive focus on efficiency improvements is unlikely to prevent catastrophic climate change. Indeed: "It is highly questionable that 12-fold to 26-fold increases in Gross World Product [over the twenty-first century, as predicted by the IPCC] are even remotely achievable because of biophysical constraints and the inability of technology to sufficiently uncouple energy and materials use from the economy."[9]

Meeting the climate change challenge depends on ending human population growth and either ending economic growth or radically transforming it, so that *some* economic growth in *some* sectors of the modern economy and in poorer countries that actually benefit from it can be accommodated without radically destabilizing Earth's climate. All the technological improvements we can muster will probably be necessary to enable this transition to a slow-growth or post-growth future—they cannot provide an alternative to it.[10] But making the necessary changes will be difficult; particularly in America, where economic growth comes close to being a sacred value.

CLIMATE CHANGE RHETORIC AND REALITY

Consider the rhetoric around popular efforts to encourage action on global climate change. Mass initiatives like Focus the Nation and Al Gore's Wecansolveit.org are morally earnest, yet cheerily optimistic. They emphasize that the climate challenge is manageable and can be met without drastically altering standards of living. In fact, they claim, climate change is chock-full of economic opportunities! In a section on their website proposing "Solutions" for a "Clean Energy Economy," Wecansolveit.org exclaims: "Thousands of new companies, millions of new jobs, and billions in revenue generated by solutions to the climate crisis—this is the clean energy economy we can adopt with today's technologies, resources, knowhow, and leadership from our elected officials." Furthermore: "A recent report showed that investment in a clean and efficient economy would lead to over three million new green-collar jobs, stimulate $1.4 trillion in new GDP, add billions in personal income and retail sales, produce $284 billion in net energy savings, all while generating sufficient returns to the U.S. Treasury to pay for itself over ten years." They conclude: "This is the opportunity of our generation—to lead the transformation to an economy that is robust without causing environmental harm."

Now, don't get me wrong. There are often good practical reasons to be optimistic in tone and emphasize the economic positives in talking to our fellow citizens. If the only way for Americans to do our part to stop global climate change is to give up our cars or keep our houses heated to 55 degrees during the winter, I don't think we'll make the effort. Fortunately, Al Gore and other optimists are right that many of the steps needed to limit greenhouse gas emissions will save us money and that the new energy technologies we need to develop and deploy can be growth industries for the U.S. economy. Similarly, estimates that the world might meet the global climate challenge by devoting only 2 percent of world GDP to the effort (as suggested in the famous 2005 Stern Report) really do give grounds for optimism.[11]

Above all, the "positive" approach recommends itself because it has led to some successes. For example, in my home state of Colorado, we have— through a direct citizens' referendum and subsequent government action— committed the state to billions of dollars of new investments in alternative energy in the next decade. We didn't achieve this by hanging crepe, but by combining moral exhortation with optimism. "Global warming is our generation's greatest environmental challenge," states Governor Bill Ritter, in his 2007 introduction to the *Colorado Climate Action Plan*. "Can Coloradans really make a difference? I believe we can, and that we have a moral obligation to try.... [Our] success depends on everyone doing his or her part. We can reduce global warming and keep our economy strong and vibrant. This is an exciting time for Colorado as we look toward an expanded New Energy Economy with new jobs, new businesses and new investments."[12] Colorado's plan is one of the most far-reaching alternative energy mandates in the United States. Score a point for the power of positive thinking.

However, this approach engenders two worries. First, it probably cannot support all the measures needed to adequately address the problem. What happens when mitigating global climate change doesn't save us money or contribute to growth, but instead costs us money or inhibits growth? Second, this approach's boosterism seems likely to further solidify the economic paradigm that is causing climate change and the rest of our environmental problems in the first place, and further entrench the economic mind-set that makes it so difficult to solve them. Can we really "expand" our economies and keep them "vibrant and strong" (i.e., growing even more) while also "reducing" global warming? Can an economy really be "robust" [that is, rapidly growing] without causing environmental harm?

The evidence suggests not. It suggests that the most important economic lesson climate change has to teach us is that the endless growth economy is unsustainable and must be replaced by a fundamentally different alternative. An economic system dedicated to generating ever more wealth and consumption for ever more people has reached its limits. It must be replaced by an economy designed to provide sufficient wealth and consumption for a finite number of people.[13] Our failure to recognize this testifies to the authority of the reigning economic orthodoxy; to the power of wishful thinking; and to our failure to grapple with the full meaning of climate change.

After all, what is global climate change? We are cooking the Earth, radically destabilizing the climate of the only home we or our descendants will have (at least for the foreseeable future). We are doing this at great risk to ourselves and great cost to the other species with whom we share this planet.

How are we creating global climate change? Not by accident, now, but consciously, as a by-product of ever more human economic activity.

Why are we creating global climate change? Because we believe—or act as though we believe—that ever more economic activity is the key to living good human lives. Or because we believe that there is no real alternative to ever more economic activity: either that it is as inevitable as Newton's laws of motion or that the alternatives are so dismal that we could never accept them. So, in responding to climate change, the assumption so far has been that we can continue to have growing economies while generating less greenhouse gas emissions, as long as we make some (perhaps difficult or costly) technological and managerial changes. But this assumption becomes less plausible with each year's reports of melting tundra and receding glaciers.

In the short term, we might make impressive strides in lowering the greenhouse gas emissions of various human activities (driving, providing electricity for our homes, etc). But there are also limits to how far we can push down these emissions: technical limits, economic limits, physical limits, and the limits associated with human inertia. Meanwhile, all else being equal, more people mean more greenhouse gas emissions and wealthier people mean more greenhouse gas emissions. The "Environmental Kuznets Curve"—in which societies, once they become sufficiently rich, start to "purchase" increased environmental protection and clean up their messes—is a very comforting idea; but studies show it does not hold for global climate change.[14] So, under the endless growth status quo, all our economic successes make ending climate change more difficult.

In the long term, it seems clear that an ever-growing economy—with more people consuming more, producing more goods and services per capita, and generating ever more wealth—will sooner or later lead to dangerous global climate change. Unless you imagine a way of life that creates *no* greenhouse gas emissions per capita, you have to accept that at some point continued growth in population, consumption, and production will generate dangerous greenhouse gas emissions and catastrophic climate change.

But OK. Let's go ahead and imagine a benign, post–fossil fuel economy where people generate no greenhouse gases at all. Even in this advanced state, shouldn't we assume that an endlessly growing human economy will create other strains on basic ecosystem services and generate other intolerable environmental costs? No one anticipated stratospheric ozone depletion as a by-product of the growing world economy; almost no one anticipated global climate change. One of global warming's main lessons should be that we are likely not at the end of this career of unanticipated consequences. Remember, according to the *Millennium Ecosystem Assessment*, currently 15 of 24 key global ecosystem services are being degraded or used unsustainably. Climate change is not the only area where humanity is pushing up against global ecological limits.

Let's further assume that we can continue to grow indefinitely and somehow avoid ecocatastrophes like ozone depletion and global climate change. Even then, ever more people consuming and producing ever more stuff will certainly require the continued taming of the Earth, the loss of wildness, and the continued displacement of nonhuman species. Some of us rank these trends right up there with climate change as ecological and moral disasters.[15] Many proposed responses to climate change, from seeding the oceans with iron to covering the Earth with solar collectors to floating vast armadas of balloons up into the stratosphere, would move humanity further toward a bland technological utopia in which we press every acre of land (and even the depths of the seas and the heights of the atmosphere) into service to our vast needs, turning Earth into an artificially controlled human life-support system, while totally displacing wild nature. This would constitute a monstrous injustice toward the rest of life on Earth—and a dangerous bet on our own foresight and technological abilities.

What all this suggests is that sooner or later (and in any case not all that much later) we are going to have to shift out of the endless growth paradigm if we hope to avoid the worst of global climate change, meet our other environmental challenges, share habitat and other resources fairly with other species, and in general act responsibly and live up to our self-proclaimed moniker *Homo sapiens*: man the wise. The good news is that planet Earth is a wonderful, richly endowed planet and human beings are resourceful and adaptable. We have good reason to believe that Earth can support a few billion people sustainably, in comfort and over the long term. But only if we aren't too greedy! Only if we accept limits to growth.

Part Seven

WHAT WE'RE FOR

WHAT WE'RE FOR

Every activist engaged in combating human-caused climate change or specific elements of the current energy economy knows that the work is primarily oppositional. It could hardly be otherwise; for citizens who care about ecological integrity, a sustainable economy, and the health of nature and people, there is plenty to oppose—burgeoning biomass logging, mountaintop-removal coal mining, inadequately regulated natural gas and oil drilling, poorly sited solar and wind developments, river-killing megadams, and new nuclear and coal plants around the globe.

These and many other fights against destructive energy projects are crucial, but they can be draining and tend to focus the conversation in negative terms. Sometimes it's useful to reframe the discourse about ecological limits and economic restructuring in positive terms, that is, in terms of *what we're for*. The following list is not comprehensive, but beauty and biodiversity are fundamentals that the energy economy must not diminish. And energy literacy, conservation, relocalization of economic systems, and family planning are necessary tools to achieve our vision of a day when resilient human communities are embedded in healthy ecosystems and all members of the land community have space enough to flourish.

In short, what we're for is leaving behind the current energy economy, which is wasteful, polluting, and centralized; assumes perpetual growth; and is anchored by nonrenewable fuels. We envision a bold leap toward a future energy economy that fosters beauty and health; that is resilient because it emphasizes renewable, community-scale energy generation; that supports durable economies, not growth; and that is informed by nature's wisdom.

Recognizing that all human economic activity is a subset of nature's economy and must not degrade its vitality is the starting point for systemic transformation of the energy system. While such a transition may seem daunting, reforms may be implemented incrementally, and the destination offers exciting possibilities for building vibrant human communities embedded in healthy ecosystems.

ENERGY LITERACY

Energy is arguably the most decisive factor in both ecosystems and human economies. It is the fulcrum of history, the enabler of all that we do. Yet few people have more than the sketchiest understanding of how energy makes the world go around.

Basic energy literacy consists of a familiarity with the laws of thermodynamics, and with the concepts of energy density and net energy (also known as energy return on energy invested, or EROEI). It requires a familiarity with the costs and benefits of our various energy sources—including oil, coal, gas, nuclear, wind, and solar. It also implies numeracy—the ability to meaningfully compare numbers referring to quantities of energy and rates of use, so as to be able to evaluate matters of scale.

Without energy literacy, citizens and policy makers are at the mercy of interest groups wanting to sell us their vision and products for the future energy economy. We hear from the fossil fuel industry, for example, that Canada's oil reserves (in the form of "tar sands") are second only to Saudi Arabia's, or that the United States has over one hundred years of natural gas thanks to newly tapped "shale gas" resources. And it's tempting to conclude (as many people do) that there are no real constraints to national fossil fuel supplies other than environmental regulations preventing the exploitation of our immense natural treasures.

On the other end of the spectrum, we hear from techno-optimists that, with the right mix of innovative energy generation and efficiency technologies, we can run the growth economy on wind, solar, hydropower, and biofuels. And it's tempting to conclude that we only need better government incentives and targeted regulatory reform to open the floodgates to a "green" high-tech sustainable future.

Energy literacy arms us with the intellectual tools to ask the right questions: What is the energy density of these new fossil fuel resources? How much energy will have to be invested to produce each energy unit of synthetic crude oil from oil shale, or electricity from thin-film solar panels? How quickly can these energy sources be brought online, and at what rate can they realistically deliver energy to consumers? When we do ask such questions, the situation suddenly looks very different. We realize that the "new" fossil fuels are actually third-rate energy sources that require immense and risky investments and may never be produced at a significant scale. We find that renewable energy technologies face their own serious constraints in energy and material needs, and that transitioning to a majority-renewable energy economy would require a phenomenal retooling of our energy and transportation infrastructure.

With energy literacy, citizens and policy makers have a basis for sound decisions. Householders can measure how much energy they use and strategize to obtain the most useful services from the smallest energy input. Cities, states, and nations can invest wisely in infrastructure to both produce and use energy with greatest efficiency and with minimal damage to the natural world. With energy literacy, we can undertake a serious, clear-eyed societal conversation about the policies and actions needed to reshape our energy system.

CONSERVATION

The current energy economy is toxic not simply because of its dependence on climate-altering fossil fuels, but also because of its massive scale and wastefulness. A first step toward reducing its global impacts is simply using less energy, a goal readily accomplished through conservation practices that are widely available and cost-effective.

Energy conservation consists of two distinct strategies: efficiency and curtailment. Energy efficiency means using less energy to produce a similar or better service. For example, we can exchange old incandescent lightbulbs for compact fluorescents or LEDs that use a fraction of the electricity and still enjoy satisfactory levels of indoor illumination. Curtailment means exactly what you'd think: cutting out a use of energy altogether. In our previous example of indoor lighting, this strategy might take the form of turning off the lights when we leave a room.

Efficiency is typically more attractive to people because it doesn't require them to change their behavior. We want services that energy provides us, not energy per se, and if we can still have all the services we want, then who cares if we're using less energy to get them? Much has been achieved with energy efficiency efforts over recent decades, but much more remains to be done: Nearly all existing buildings need to be better insulated, and most electric power plants are operating at comparatively dismal efficiencies, to mention just two examples.

Unfortunately, increasing investments in energy efficiency typically yield diminishing returns. Initial improvements tend to be easy and cheap; later ones are more costly. Sometimes the energy costs of retooling or replacing equipment and infrastructure wipe out gains from efficiency. Nevertheless, the early steps toward efficiency are almost always rewarding.

While curtailment of energy use is a less inviting idea, it offers clearer savings. By simply driving fewer miles we unequivocally save energy, whether our car is a more or less efficient model. We've gotten used to using electricity and fuels to do many things that can be done well enough with muscle power, or that don't need doing at all.

Conservation helps us appreciate the energy we use. It fosters respect for resources, and for the energy and labor that are embodied in manufactured products. It reduces damage to already stressed ecosystems and helps us focus our attention on dimensions of life other than sheer consumption.

During the latter decades of the twentieth century, most Americans achieved a standard of living that was lavish from both historical and cross-cultural perspectives. They were coaxed and cajoled from cradle to grave by advertising to consume as much as possible. Simply by reversing the message of this incessant propaganda, people might be persuaded to make do with less—as occurred during World War II—and be happier as well. Many social scientists claim that our consumptive lifestyle damages communities, families, and individual self-esteem. A national or global ethic of conservation could even be socially therapeutic.

RESILIENCE

Resilience is "the capacity of a system to withstand disturbance while still retaining its fundamental structure, function, and internal feedbacks." Resilience contrasts with *brittleness*—the tendency to shatter and lose functionality when impacted or perturbed.

Ecologists who study resilience in natural systems have noted that ecosystems tend to progress through a series of phases: growth, consolidation and conservation, release (or "collapse"), and reorganization. Each turning of this adaptive cycle provides opportunities for individual species and whole systems to innovate in response to external and internal change (i.e., disturbance). Resilient ecosystems (in the early *growth* phase) are characterized by species diversity; many of the organisms within such systems are flexible generalists, and the system as a whole contains multiple redundancies. In contrast, less resilient ecosystems tend to be more brittle, showing less diversity and greater specialization particularly in the *consolidation* phase.

Resilience can be applied to human systems as well. Our economic systems, in particular, often face a trade-off between resilience and efficiency. Economic efficiency implies specialization and the elimination of both inventories and redundancy (which typically guarantee greater resilience). If a product can be made most cheaply in one region or nation, manufacturing is concentrated there, reducing costs to both producers and consumers. However, if that nation were to suddenly find it impossible to make or ship the product, that product would become unavailable everywhere. Maintaining dispersed production and local inventories promotes availability under crisis conditions, though at the sacrifice of economic efficiency (and profits) in "normal" times.

From a resilience perspective one of the most vulnerable human systems today is the American transportation system. For over seventy years we've spent trillions of dollars building transportation infrastructure that is completely dependent (i.e., "specialized") on affordable petroleum fuels, and we've removed or neglected most alternative methods of transport. As petroleum fuels become less affordable, the effects reverberate throughout the system.

Resilience becomes more of a priority during periods of crisis and volatility, such as the world is experiencing today. Households, towns, and regions are better prepared to endure a natural disaster such as a flood or earthquake if they have stores of food and water on hand and if their members have a range of practical self-sufficiency skills.

While the loss of economic efficiency implies trade-offs, resilience brings incidental benefits. With increased local self-sufficiency comes a shared sense of confidence in the community's ability to adapt and endure. For the foreseeable future, as global energy, finance, and transport systems become less reliable, the rebalancing of community priorities should generally weigh in favor of resilience.

ECO-LOCALISM

A central strategy needed to increase societal resilience is *localization*—or, perhaps more accurately, *re*localization. Most pre-industrial human societies produced basic necessities locally. Trade typically centered on easily transportable luxury goods. Crop failures and other disasters therefore tended to be limited in scope: If one town was devastated, others were spared because they had their own regional sources—and stores—of necessities.

Economic globalization may have begun centuries ago with the European colonization of the rest of the world, but it really took hold during the past half century with the advent of satellite communications and container ships. The goal was to maximize economic growth by exploiting efficiency gains from local specialization and global transport. In addition to driving down labor costs and yielding profits for international corporations, globalization maximized resource depletion and pollution, simplified ecosystems, and eroded local systems resilience.

As transport fuel becomes less affordable, a return to a more localized economic order is likely, if not inevitable. The market's methods of rebalancing economic organization, however, could well be brutal as global transport networks become less reliable, transport costs increase, and regions adapt to less access to goods now produced thousands of miles away.

Government planning and leadership could result in a more organized and less chaotic path of adaptation. Nations can begin now to prioritize and create incentives for the local production of food, energy, and manufactured products, and the local development of currency, governance, and culture.

Natural ecological boundaries—such as watersheds—bordered traditional societies. Bioregions defined by waterways and mountain ridges could thus become the basis for future relocalized economic and political organization.

Deliberate efforts to relocalize economies will succeed best if the benefits of localism are touted and maximized. With decentralized political organization comes greater opportunity for participation in decision making. Regional economic organization offers a wide variety of productive local jobs. Society assumes a human scale in which individuals have a sense of being able to understand and influence the systems that govern their lives. People in locally organized societies see the immediate consequences of their production and waste disposal practices, and are therefore less likely to adopt an "out of sight, out of mind" attitude toward resource depletion and pollution. Local economic organization tends to yield art, music, stories, and literature that reflect the ecological uniqueness of place—and local culture in turn binds together individuals, families, and communities, fostering a sense of responsibility to care for one another and for the land.

BEAUTY

Discussions about energy rarely focus on beauty. But the presence or absence of this ineffable quality offers us continual clues as to whether or not society is on a regenerative and sustainable path, or on the road to further degrading nature.

From the time of the earliest cave paintings, human ideals of beauty have been drawn from the wild world. Animals, plants, rivers, oceans, and mountains all tend to trigger a psychological response describable as pleasure, awe, and wonder. The sight of a great tree or the song of a goldfinch can send poets and mystics into ecstasy, while the deep order inherent in nature inspires mathematicians and physicists.

Nature achieves its aesthetic impact largely through anarchic means. Each part appears free to follow its own inner drives, exhibiting economy, balance, color, proportion, and symmetry in the process. And all of these self-actualizing parts appear to cooperate, with multiple balancing feedback loops maintaining homeostasis within constantly shifting population levels and environmental parameters. The result is beauty.

Ugliness, by contrast, is our unpleasant aesthetic response to the perception that an underlying natural order has been corrupted and unbalanced—that something is dreadfully out of place.

Beauty is a psychological and spiritual need. We seek it everywhere and wither without it. We need beauty not as an add-on feature to manufactured products, but as an integral aspect of our lives.

With the gradual expansion of trade—a process that began millennia ago but that quickened dramatically during the past century—beauty has increasingly become a valuable commodity. Wealthy patrons pay fortunes for rare artworks, while music, fashion, architecture, and industrial design have become multibillion-dollar industries. Nature produces the most profound, magnificent, and nurturing examples of beauty in endless abundance, for free.

Industrialism, resulting from high rates of energy use, tends to breed ugliness. Our ears are bombarded by the noise of automobiles and trucks to the point that we can scarcely hear birdsong. The visual blight of highways, strip malls, and box stores obscures natural vistas. With industrial-scale production of buildings, we have adopted standardized materials produced globally to substitute for local, natural materials that fit with their surroundings. But industrialism does not just replace and obscure natural beauty—it actively destroys it, gobbling up rivers and forests to provide resources for production and consumption.

Large-scale energy production—whether from coal mines and power plants, oil derricks and refineries, or massive wind and solar installations—comes at a cost of beauty. While some energy sources are inherently uglier than others, even the most benign intrude, dominate, and deplete if scaled up to provide energy in the quantities currently used in highly industrialized nations.

The aesthetic impact of industrial processes can be mitigated somewhat with better design practices. But the surest path to restoring the beauty of nature is to reduce the scale of human population and per capita production and consumption. Returning to a sustainable way of life need not be thought of as sacrifice; instead it can be seen as an opportunity to increase aesthetic pleasure and the spiritual nourishment that comes from living in the midst of incalculable beauty.

BIODIVERSITY

The family of life on Earth is large: More than a million species have been identified and formally described by taxonomists, and estimates of the total number of species on the planet range from 3 million to 100 million. We humans depend for our very existence on this web of life of which we are a part. Indeed, it is part of us: Each human is inhabited by thousands of species of microbes that enable digestion and other basic functions. Yet through our species' appropriation and destruction of natural habitat we are shredding microbial, forest, prairie, oceanic, riparian, desert, and other ecosystems. Habitat loss, overharvesting, climate change, and other results of human numbers and behavior endanger untold numbers of species with extinction.

Extinction is nothing new: It is an essential part of the process of evolution. Throughout the billions of years of life's history, life forms have appeared, persisted for thousands or millions of years, and vanished, usually individually but occasionally in convulsive mass events triggered by geological or astrophysical phenomena. There were five ancient extinction events so catastrophic that 50–95 percent of all species died out.

Today humans are bringing about the sixth mass extinction in the history of life on Earth. While the normal rate of extinction is about one in a million species per year, the extinction rate today is roughly a thousand times that. According to recent studies, one in five plant species faces extinction as a result of climate change, deforestation, and urban growth. One of every eight bird species will likely be extinct by the end of this century, while one-third of amphibian and one-quarter of mammal species are threatened.

As species disappear, we are only beginning to understand what we are losing. A recent United Nations study determined that businesses and insurance companies now see biodiversity loss as presenting a greater risk of financial loss than terrorism—a problem that governments currently spend hundreds of billions of dollars per year to contain or prevent.

Nonhuman species perform ecosystem services that only indirectly benefit our kind, but in ways that often turn out to be crucial. Phytoplankton, for example, are not a direct food source for people, but comprise the base of oceanic food chains, in addition to supplying half of the oxygen produced each year by nature. The abundance of plankton in the world's oceans has declined 40 percent since 1950, according to a recent study, for reasons not entirely clear. This is one of the main explanations for a gradual decline in atmospheric oxygen levels recorded worldwide.

Efforts to determine a price for the world's environmental assets have concluded that the annual destruction of rainforests alone entails an ultimate cost to society of $4.5 trillion—roughly $650 for each person on the planet. Many species have existing or potential economically significant uses, but the value of biodiversity transcends economics: The spiritual and psychological benefits to humans of interaction with other species are profound.

Most fundamentally, however, nonhuman species have intrinsic value. Shaped by the same forces that produced humanity, our kin in the community of life exist for their own sake, not for the pleasure or profit of people. It is the greatest moral blot, the greatest shame on our species, for our actions to be driving other life forms into the endless night of extinction.

FAMILY PLANNING

The human demographic explosion, amplified by rapacious consumption in the overdeveloped world, is at the root of the global eco-social crisis. Virtually every environmental and social problem is worsened by over-population. With more mouths to feed—and freshwater becoming scarcer and topsoil eroding—global famine becomes an ever-greater likelihood. An expanding population leads to increased consumption of just about every significant resource, and thus to increasing rates of ecological damage, from deforestation to climate change.

Family planning helps avert those threats. If we want future generations to enjoy a healthy planet with wild spaces, biodiversity, abundant resources, and a livable climate we should reduce fertility now.

But family planning can do more than mitigate future resource depletion; it has direct and in some cases nearly immediate benefits. Some of those benefits are economic. For example, Ireland's declining birth rate in the 1970s is often credited as one of the factors leading to its economic boom in the 1980s and 1990s. China's one-child policy similarly contributed to its economic ascendancy. The mechanism? In poor societies where family size is typically large, all household income must go toward food and shelter, and none is left over for education and business formation. If the birth rate is reduced, household income is freed up to improve quality of life and economic prospects for the next generation.

Without access to contraceptives, the average woman would have from 12 to 15 pregnancies in her lifetime. In contrast, women in industrial nations want, on average, only two children.

It turns out that when women are economically and—this is critical—*culturally* empowered to make decisions about their own fertility, the result is improved health for mother and children, fewer unplanned pregnancies and births, and reduced incidence of abortion. Numerous studies have shown that women who have control over their fertility also tend to have more educational and employment opportunities, enhancing their social and economic status and improving the well-being of their families.

AFTERWORD

Places Where the Wind Carries the Ashes of Ancestors

LISI KRALL

We can be sympathetic to Adam Smith's preoccupation with the wealth of nations. After all, if we were still in the undeveloped world economy of the late eighteenth century, with less than a billion people, his preoccupation might be understandable. We now stand in a different economic world, a world dominated by exponential growth curves, a "full world" in Herman Daly's vernacular. Energy consumption and production, population, carbon dioxide emissions, biodiversity loss, soil depletion, and many more barometers of our imbalance all grow explosively. Perhaps poverty too is growing exponentially, since there are more impoverished people on the planet now than were alive at the time that Adam Smith was writing his tome on capitalism. Clearly our economic experiment with increasing the wealth of nations has not been an unmitigated success.

We live in a mature capitalist economy where the dynamic and logic as well as the internal contradictions of our economic system are fully developed and revealed. The logic and imperative of growth, punctuated with cyclical and secular stagnation, inequality, and unemployment define our economic landscape. Our method for solving these systemic problems is to encourage growth. Yet at the same time we have hit a biophysical wall. Clearly we confront economic contradictions with no easy resolution.

It isn't surprising that with this daunting situation we focus our hope on the technological possibilities of green energy and the belief that well-informed markets will provide it. We understand that there is an intimate dance between market capitalism and energy; clearly they are inextricably connected. The problem is that the cheap availability of fossil fuel which accommodated the full flowering of industrial capitalism has given us a distorted picture of what is possible and has set us on an unsustainable path. In order to keep our immense economic machine moving, the historical oddity of abundant cheap energy has now become an economic necessity and an expectation for the future.

The reality is this: We are making the transition out of fossil fuel later than we should, because *the vested interests* find this the best course for business—but it is clearly not the best course for society. The interest of business is to extract every morsel of fossil fuel that can be economically retrieved, no matter what the ecological consequences. And we should remind ourselves that it isn't clear that green energy will satisfactorily fill the void. In the final chapter of our use of fossil fuels we will commit ourselves to the messiest, dirtiest, most socially and economically disruptive transition we can possibly muster unless we are willing to change course. We now retreat behind our dominant ideologies—the promise of the entrepreneurial spirit and technological prowess—for guidance, and we fail to see that these are no longer the solutions to the problems of our civilization. They are palliatives at best.

We might consider whether we are simply in need of closing the final chapter of our grand experiment with excess and the relentless domestication of the planet that requires ever-increasing amounts of energy. This would necessitate something more fundamental than a technological energy fix forged by entrepreneurial motivations. Perhaps the real question of progress is not how to forge a new energy frontier, but how to forge a different model of economic organization and purpose, a model that isn't predicated on never-ending growth and a belief that there are no real biophysical limits.

❖

LAST SUMMER MY FATHER—son of immigrant and homesteader parents, World War II and Korean War veteran and patriot—died peacefully surrounded by his family. His wish was that we should spread his ashes from the top of Sheep Mountain, a place about ten miles north of the coal-mining town in southwestern Wyoming where Dad and most of our clan came into this world and many still live. After my father's memorial service, my extended family gathered in the parking lot of the Jubilee grocery store to load into the available vehicles and make the pilgrimage to the top of this sacred mountain. It is a ridge sitting at about 10,000 feet where the wind could be relied on to carry the ashes of my father. The top is bare, craggy, and weathered, and a few ancient and resilient limber pine hang on the edge that drops off into the vast Wyoming landscape. The exposed sandstone, the remnants of ancient seas, is covered with orange, gray, and black lichens. Nothing survives here without steadfast purpose.

I can look out and see for hundreds of miles into the high desert steppe of southwestern Wyoming and the places my father roamed for over ninety years: Whiskey Basin, Cow Hollow, the Fontenelle Basin, the Hamsfork drainage, the Wind Rivers off in the distance. These are the places that defined his life, and he always felt privileged by his destiny. But this place is also part of the Wyoming overthrust belt and so, looking out, I can see the once great expanse of wildness dotted with gas fields. The air is filled not with the clarity of the past but with the dust and ozone of the present. The empty spaces are now punctuated with the indelible markings of our energy diet. And not simply gas and oil wells and all of their paraphernalia, but windmills too —they're popping up everywhere, further altering the once empty horizon.

As we stand atop the mountain I am told by my nephew, Phil, that there is a plan to bring a massive power line over the top of Sheep Mountain. In the push for energy this sacred ground will be sacrificed. The world is littered with the sadness of such choices. People everywhere, the Ogoni of the Nigerian Delta, the Aysen People of Patagonia, the Aihuar of the Ecuadorian Amazon, the Mashpee Wampanoag of Cape Cod, the Aquinnah Wampanoag of Martha's Vineyard, the Marcellus Shale People of Central New York, the Mountaintop-Removal People of West Virginia—and us too, the High Steppe People of Southwestern Wyoming—we all experience the violation of our sacred places. The list goes on and on. The cost-benefit analyses indicate this is our best option. I am sick to death of it.

Here where the wind blows, green energy will not stop the process of this violation—because to get the energy of the wind out of here will take great disruption, just as taking the gas and oil and coal have. In this full world I think maybe we should set aside our cost-benefit analyses and simply stop the violation of sacred places.

Herman Daly invoked the words of John Stuart Mill a long time ago when he started to think about the wisdom of ending economic growth on a biophysically limited planet. Mill was confused about many economic questions, but he at least understood how to pose the important ones. Here is one that hit the nail on the head: "If the earth must lose the great portion of its pleasantness which it owes to things that the unlimited increase of wealth and population would extirpate from it, for the mere purpose of enabling it to support a larger, but not a happier or better population, I sincerely hope, for the sake of posterity, that they will be content to be stationary, long before necessity compels them to it." Energy will eventually compel us to be stationary but we should take Mill's words to heart. Perhaps it would be wise to stop growing long before we are forced to do so.

There is a collective voice gathering force, wisdom forged out of the connection between humans and the particular places they inhabit and know to be sacred; a wisdom in contradiction with the economic forces of our time. It is necessary to find a way to let this wisdom challenge our present economic arrangements. The real question of progress is whether we can construct an economy that doesn't defile the sacred; the resting places of the ashes and memories of our ancestors. This isn't a question of economic value. Instead it is a question of the value of embedding ourselves in the all-sustaining and resonant magic of the Earth. It is a question of the value of greater purpose and less human hubris. It is a question of the value of appreciating that it isn't all about us. Our energy challenge should be the catalyst to step back from the brink of this abyss before the waning hours of the fossil fuel economy alters our sense of place so irrevocably that we no longer see ourselves as Earthlings.

ACKNOWLEDGMENTS

No single volume can fully cover the myriad issues that encompass humanity's energy predicament. And no (reasonably sized) list can comprehensively acknowledge the people who have influenced our thinking about energy policy, and who have directly lent support to this project. We thank them all, named and unnamed.

First and foremost we are grateful to our extraordinary collaborators at Post Carbon Institute—Asher Miller and Richard Heinberg. ENERGY, *The Energy Reader,* and the associated Energy Reality campaign simply could not have happened without their wisdom, knowledge, and dedicated labors. Also indispensable to the effort were Dan Imhoff and Jane Mackay at Watershed Media, the small nonprofit publisher with a big vision about sustainability and the fate of the Earth. We thank our colleagues Debbie Ryker, Esther Li, Fay Li, Lorren Butterwick, and Nadine Lehner at the Foundation for Deep Ecology and Conservacion Patagonica, and most especially acknowledge the incredible leadership and inspiration provided by Doug and Kris Tompkins. They make the world a wilder, more beautiful place.

In addition to the contributing writers in this book, many activists and thinkers on energy and related topics influenced our thinking or provided direct assistance, including Jerry Mander, John Davis, Katie Fite, Howard Wilshire, Aaron Sanger, Helena Norberg-Hodge, Robert Rapier, Tony Clarke, Stan Moore, David Pimentel, Michael Bomford, Vandana Shiva, John Bellamy Foster, Fred Magdoff, Thomas Power, and Thomas Pringle. Solar Bus guru Gary Beckwith, publisher Jack Shoemaker, Clay Stranger at Rocky Mountain Institute, and Duncan McCallum were helpful with production matters. Kevin Cross produced the graphics in Part Three. In typically meticulous fashion, typesetter Timothy Rice, proofreader Mary Elder Jacobsen, and indexer Leonard Rosenbaum produced outstanding work for this Reader edition of the book.

Finally, we offer our thanks to all the writers who contributed work to the book and to all the citizens who may use it to inform their activism, as we work to create a future energy economy that sustains beauty, biodiversity, and flourishing human communities.

—*Tom Butler, Daniel Lerch, and George Wuerthner*

CREDITS

The editors are grateful to all of the photographers whose work appears in this book and its large-format version. We offer special thanks to the individual who took the cover shot of the Deepwater Horizon drilling platform aflame in the Gulf of Mexico. Although we spent countless hours trying to locate the individual who captured that dramatic image and were prepared to pay the going rate for the photo's use (and, of course, still are), he or she chose to remain anonymous. We can only presume that the photographer was either an oil company employee or a first responder who felt his or her job precluded being publicly associated with the accident photo.

The overarching goal of ENERGY and The Energy Reader is to bring into focus the destructive nature of the extractive economy that has produced the eco-social crisis, characterized by diminishing beauty, biodiversity, and health across the planet. In our view the cover photo captures that destructive nature in spectacular fashion, and we chose to use it for noncommercial purposes as part of the book and its related energy literacy campaign to spur social and environmental activism. We thank this unnamed photographer for his or her contribution to waking up society to the crisis in which we are all ensnared.

Other photos: Page *xviii*, Lawrence Parent; 22, EcoFlight; 102, J. Henry Fair; 150, Getty Images/Jacques Jangoux; 214, Mark Gocke; 262, Doug Tompkins; 318, iStockphoto.com.

CONTRIBUTORS

RICHARD BELL is an author, editor, and political consultant. After working with activists opposing a nuclear power plant in Seabrook, New Hampshire, Richard coauthored the original 1982 edition of *Nukespeak: Nuclear Language, Myths, and Mindset*; the book was updated and reissued in 2011. He later served as research director for the Democratic Senatorial Campaign Committee, new media director at the Democratic National Committee, and vice president for communications at the Worldwatch Institute.

WENDELL BERRY is an American poet, novelist, essayist, conservationist, philosopher, visionary, and farmer. His books include *The Unsettling of America, Jayber Crow, Life Is a Miracle, The Mad Farmer Poems*, and *Bringing It to the Table: On Farming and Food*. He is an elected member of the Fellowship of Southern Writers and a recipient of The National Humanities Medal, among many other honors.

SHEILA BOWERS is a citizen activist with solardoneright.org. For several years she has been researching the economic, political, and legal biases that promote industrial-scale energy development while artificially impeding the growth of environmentally sound distributed generation.

LESTER BROWN is founder and president of the Earth Policy Institute, and is the author or coauthor of some fifty books. He is the recipient of many prizes and awards including 25 honorary degrees, a MacArthur Fellowship, the United Nations' Environment Prize, the World Wide Fund for Nature Gold Medal, the Blue Planet Prize, and the Borgström Prize awarded by the Royal Swedish Academy of Agriculture and Forestry. He founded the Worldwatch Institute in 1974 and launched its respected series of research papers and annual reports on the state of the global environment.

MARTIN CURRY is a Saginaw Chippewa tribal member; he lives in LaPointe, Wisconsin, and is a caregiver for his father.

DAVID EHRENFELD is a professor at Rutgers University and a leading conservation biologist. He was the founding editor of the journal *Conservation Biology*, was a longtime columnist for *Orion*, and is the author of several books, including *Becoming Good Ancestors: How We Balance Nature, Community, and Technology* and *The Arrogance of Humanism*.

ETC GROUP is an international civil society organization that promotes critical analysis of emerging technologies. Based in Canada, the organization works to sustain cultural and biological diversity and to advance food sovereignty and human rights. This essay was adapted from the report *Retooling the Planet*, which ETC Group prepared for the Swedish Society for Nature Conservation; lead authors were Diana Bronson, Pat Mooney, and Kathy Jo Wetter.

GLORIA FLORA works for public land sustainability through her organization Sustainable Obtainable Solutions, focusing on large landscape conservation, state-based climate change solutions, and promoting the production and use of biochar. She is a fellow of the Post Carbon Institute and of the Center for Natural Resources and Environmental Policy. In her twenty-three years with the U.S. Forest Service she served as the Forest Supervisor for two national forests. She has won multiple awards for her leadership, courage, and environmental stewardship.

DAVID FRIDLEY has been a staff scientist in the Energy Analysis Program at the Lawrence Berkeley National Laboratory in California since 1995. He has more than thirty years of experience working and living in China in the energy sector, and is a fluent Mandarin speaker. He spent twelve years working in the petroleum industry both as a consultant on downstream oil markets and as business development manager for Caltex China. He has written and spoken extensively on the energy and ecological limits of biofuels, and he is a Fellow of Post Carbon Institute.

JEFF GOODELL is a longtime contributing editor at *Rolling Stone*, has written for many periodicals. His books include *Big Coal: The Dirty Secret Behind America's Energy Future* and *How to Cool the Planet: Geoengineering and the Audacious Quest to Fix Earth's Climate*.

JOHN MICHAEL GREER is the Grand Archdruid of the Ancient Order of Druids in America, a peak-oil theorist, and the author of more than 20 books on a wide range of subjects. His recent books include *The Long Descent: A User's Guide to the End of the Industrial Age*, *The Ecotechnic Future: Exploring a Post-Peak World*, *The Wealth of Nature: Economics as if Survival Mattered*, and *Apocalypse Not*. Greer lives in Cumberland, Maryland.

CHARLES A. S. HALL is a systems ecologist with a focus on energy and is ESF College Foundation Distinguished Professor at the State University of New York in the College of Environmental Science & Forestry (ESF). He has held positions at the Brookhaven Laboratory, Cornell University, and other institutions, and is the author of more than 250 publications.

JAMES HANSEN is director of NASA's Goddard Institute for Space Studies in New York and a leading atmospheric scientist. An internationally recognized authority on climate change, he has become one of the most prominent public communicators about the dangers of climate disruption caused by human activity.

RICHARD HEINBERG is senior fellow-in-residence at the Post Carbon Institute and is widely regarded as one of the world's foremost peak oil educators. He has written scores of essays and articles for a wide range of popular and academic periodicals, and he is the author of ten books including *The End of Growth: Adapting to Our New Economic Reality*, *Powerdown: Options and Actions for a Post-Carbon World*, and *The Party's Over: Oil, War and the Fate of Industrial Societies*.

BRIAN L. HOREJSI is a wildlife scientist with a PhD from the University of Calgary and a BS in forestry from the University of Montana. Formerly a range management forester for the Alberta Forest Service, he has conducted studies of grizzly and black bears, bighorn sheep, moose, and caribou, and for more than two decades he has been a conservation biology consultant and activist focused on preserving biological diversity. He lives in the Rocky Mountain foothills.

DAVID HUGHES is a geoscientist who has studied the energy resources of Canada for nearly four decades, including thirty-two years with the Geological Survey of Canada as a scientist and research manager. Over the past decade he has researched, published, and lectured widely on global energy and sustainability issues in North America and internationally. He is a board member of the Association for the Study of Peak Oil and Gas–Canada and is a fellow of the Post Carbon Institute.

WES JACKSON is a plant geneticist and one of the foremost thinkers in sustainable agriculture. In 1976 he founded The Land Institute to develop "natural systems agriculture." Jackson's many honors include being named a Pew Conservation Scholar and a MacArthur Fellow. He received the Right Livelihood Award in 2000, and he is a Fellow of Post Carbon Institute. His books include *New Roots for Agriculture*, *Becoming Native to This Place*, and *Consulting the Genius of the Place*, from which this essay is adapted.

BOB KING is an engineer with decades of experience developing small-scale hydroelectric, solar, and wind power projects. He is the president of Ashuelot River Hydro in Keene, New Hampshire. He owns and operates hydroelectric dams in New England and upstate New York, and he is deeply engaged in land conservation and renewable energy activism.

LISI KRALL is a Professor of Economics at the State University of New York at Cortland where she teaches Political Economy and Environmental and Ecological Economics. As a heterodox economist and an interdisciplinary scholar she publishes work in a wide range of journals, from *Conservation Biology* to the *Cambridge Journal of Economics*. Her book *Proving Up: Domesticating Land in U.S. History* uses institutional economics to explore the interconnections of economy, culture, and land in the American experience.

WINONA LADUKE is an enrolled member of the Mississippi Band Anishinaabeg who lives and works on the White Earth Reservations. She is an internationally renowned Native American Indian activist and advocate for environmental, women's, and children's rights. A graduate of Harvard and Antioch Universities with advanced degrees in rural economic development, she is founder and Executive Director of Honor the Earth, a national advocacy group encouraging public support and funding for native environmental groups.

HARVEY LOCKE A native of Alberta, Canada, is a founder and Strategic Advisor of the Yellowstone to Yukon Conservation Initiative. Globally known for his work on wilderness, national parks, and large landscape conservation, he is Strategic Conservation Advisor at the WILD Foundation, Senior Advisor for Conservation at the Canadian Parks and Wilderness Society, a member of the IUCN World Commission on Protected Areas, and a leading thinker on integrating nature conservation and climate change policy.

AMORY LOVINS is an energy visionary and the author of hundreds of scientific papers and 31 books, the latest of which is the 2011 "grand synthesis" *Reinventing Fire: Bold Business Solutions for the New Energy Era.* Lovins cofounded and chairs Rocky Mountain Institute, an independent, nonpartisan, nonprofit think-and-do tank that collaborates with the private sector to drive the efficient and restorative use of resources.

SANDRA B. LUBARSKY is a writer and professor whose research interests focus on the intersection of aesthetics and sustainability. She founded the master's program in sustainable communities at Northern Arizona University and currently directs the Sustainable Development program at Appalachian State University in North Carolina.

BILL MCKIBBEN is the author of a dozen books about the environment, including *The End of Nature* (1989), the first book for a general audience on climate change, and *Eaarth: Making a Life on a Tough New Planet* (2010). He is a founder of the grassroots climate campaign 350.org, which has coordinated 15,000 rallies in 189 countries since 2009. A scholar-in-residence at Middlebury College, he holds honorary degrees from a dozen colleges and is a fellow of Post Carbon Institute. In 2011 he was elected a fellow of the American Academy of Arts and Sciences.

ERIK MOLVAR is a wildlife biologist and executive director of Biodiversity Conservation Alliance in Laramie, Wyoming. He is the author of numerous guidebooks including *Wild Wyoming, Hiking Glacier and Waterton Lakes National Parks, Alaska on Foot: Wilderness Techniques for the Far North,* and *Wyoming's Red Desert: A Photographic Journey.*

DAVID MURPHY is an assistant professor within the Department of Geography and an associate of the Institute for the Study of the Environment, Sustainability, and Energy at Northern Illinois University. His research focuses on the role of energy in economic growth, with a specific focus on net energy.

JUAN PABLO ORREGO is one of Chile's foremost conservationists and a recipient of the Goldman Environmental Prize and the Right Livelihood Award. He is founder and director of Ecosistemas. In the 1990s he was a leader in the campaign against six large dams proposed for southern Chile's Biobío River. He is presently helping lead an international coalition of conservationists fighting a megadams scheme that would destroy wild rivers in Chilean Patagonia.

BILL POWERS is the principal of Powers Engineering, an air-quality-consulting engineering firm established in San Diego in 1994. He is a respected analyst on issues relating to electrical transmission, power plant emissions, and permitting.

VANDANA SHIVA is a world-renowned environmental leader and thinker. Director of the Research Foundation on Science, Technology, and Ecology, she is the author of many books, including *Stolen Harvest: The Hijacking of the Global Food Supply* (2000) and *Soil Not Oil: Environmental Justice in an Age of Climate Crisis* (2008). She is the founder of Navdanya, a movement promoting diversity and use of native seeds, and a recipient of the Right Livelihood Award. She holds a master's degree in the philosophy of science and a PhD in particle physics.

RACHEL SMOLKER is a codirector of Biofuelwatch. She has researched, written about, and organized extensively on the threats to forests, biodiversity, people, and the climate from biofuels. She has a PhD in ecology from the University of Michigan, worked previously as a field biologist, and is author of *To Touch a Wild Dolphin.*

SANDRA STEINGRABER is an ecologist, cancer survivor, and acclaimed writer. She is also an internationally recognized authority on the environmental links to cancer and human health. She is a scholar in residence at Ithaca College, a columnist for *Orion* magazine, and the author of *Living Downstream* and *Having Faith: An Ecologist's Journey to Motherhood*.

MICHAEL WATTS is Director of African Studies at the University of California, Berkeley. The author of eight books, he has published widely on Nigeria and the Niger Delta over the last three decades. He was awarded a Guggenheim Fellowship in 2001 for his work on the impact of oil in Africa.

R. JAMES WOOLSEY is a venture partner in the New York venture capital firm Lux Capital, a Senior Fellow at Yale University's Jackson Institute for Global Affairs, and the Chairman of the Foundation for Defense of Democracies. A national security and energy expert, he held presidential appointments in two Democratic and two Republican administrations, including as Director of Central Intelligence.

GEORGE WUERTHNER is the Ecological Projects Director for the Foundation for Deep Ecology. He has published 35 books on a wide variety of conservation topics, including *Wildfire: A Century of Failed Forest Policy* and *Welfare Ranching: The Subsidized Destruction of the American West*.

ENDNOTES

PART TWO THE PREDICAMENT

Richard Heinberg

1. See the author's discussion of peaking date forecasts in Richard Heinberg, *The Party's Over: Oil, War and the Fate of Industrial Societies* (Gabriola Island, BC: New Society Publishers, 2005, 2nd ed.), 111–118.

2. See, for example, Peter Jackson, *The Future of Global Oil Supply: Understanding the Building Blocks* (Englewood, CO: IHS CERA, 2009), http://www.scribd.com/doc/22666201/The-Future-of-Global-Oil-Supply.

3. Robert Hirsch, Roger Bezdek, and Robert Wendling, *Peaking of World Oil Production: Impacts, Mitigation, & Risk Management* (McLean, VA; SAIC, 2005), http://www.netl.doe.gov/publications/others/pdf/oil_peaking_netl .pdf.

4. David Pimentel and Marcia Pimentel, "The Future of American Agriculture," in *Sustainable Food Systems*, ed. Dietrich Knorr (Roslyn, NY: AVI Publishing Co., 1983).

5. Rich Pirog and Andrew Benjamin, *Calculating Food Miles for a Multiple Ingredient Food Product* (Ames, IA: Leopold Center for Sustainable Agriculture, 2005).

6. See, for example, *Crude Oil: Uncertainty about Future Oil Supply Makes It Important to Develop a Strategy for Addressing a Peak and Decline in Oil Production* (Washington, DC: US Government Accountability Office, February 2007), http://www.gao.gov/new.items/d07283.pdf.

7. See, for example, Daniel Lerch, *Post Carbon Cities: Planning for Energy and Climate Uncertainty* (Sebastopol, CA: Post Carbon Institute, 2007).

8. Richard Heinberg, *The Oil Depletion Protocol: A Plan to Avert Oil Wars, Terrorism and Economic Collapse* (Gabriola Island, BC: New Society Publishers, 2006).

Charles A. S. Hall

1. See Charles A. S. Hall, "Introduction to Special Issue on New Studies in EROI (Energy Return on Investment)," *Sustainability* 3, no. 10 (October 2011), 1773–1777; see also Charles A. S. Hall and Kent Klitgaard, *Energy and the Wealth of Nations: Understanding the Biophysical Economy* (New York: Springer, 2011), and sources therein. The term "Energy returned on energy invested" (EROEI) is also commonly used.

2. Carey W. King and Charles A. S. Hall, "Relating Financial and Energy Return on Investment," *Sustainability* 3, no. 10 (October 2011), 1810–1832.

3. David J. Murphy et al., "Order from Chaos: A Preliminary Protocol for Determining the EROI of Fuels," *Sustainability* 3, no. 10 (October 2011), 1888–1907.

4. King and Hall, "Relating Financial and Energy Return on Investment."

5. Clark W. Bullard, Bruce Hannon, and Robert Herendeen, *Energy Flow through the US Economy* (Champaign, IL: University of Illinois Press, 1975); Bruce Hannon, *Analysis of the Energy Cost of Economic Activities: 1963–2000*, Energy Research Group Document No. 316 (Champaign, IL: University of Illinois, 1981); Robert Herendeen and Clark Bullard, "The Energy Costs of Goods and Services," *Energy Policy* 3 (1975), 268; Robert Costanza, "Embodied Energy and Economic Valuation," *Science* 210 (December 12, 1980), 1219–1224.

6. Robert Kaufmann and Charles Hall, "The Energy Return on Investment of Imported Petroleum," in *Energy and Ecological Modeling*, ed. W. J. Mitsch et al. (Amsterdam: Elsevier Scientific, 1981), 697–701. See also chapter 8 of Charles A. S. Hall, Cutler Cleveland, and Robert Kaufmann, *Energy and Resource Quality: The Ecology of the Economic Process* (New York: Wiley, 1986).

7. Howard T. Odum, "Energy, Ecology and Economics," *AMBIO* 2 (1973), 220–227.

8. Charles A. S. Hall, "Migration and Metabolism in a Temperate Stream Ecosystem," *Ecology* 53 (1972), 585–604.

9. See Richard Heinberg, *Searching for a Miracle: Net Energy Limits and the Fate of Industrial Society* (San Francisco: International Forum on Globalization and the Post Carbon Institute, 2009), http://www.postcarbon.org /report/44377-searching-for-a-miracle.

David Fridley

1. John Whims, *Pipeline Considerations for Ethanol*, Kansas State University Department of Agricultural Economics (Manhattan, KS: Kansas State University, August 2002), http://www.agmrc.org/media/cms /ksupipelineethl_8BA5CDF1FD179.pdf.

2. Charles A.S. Hall, Robert Powers, and William Schoenberg, "Peak Oil, EROI, Investments and the Economy in an Uncertain Future," in *Biofuels, Solar and Wind as Renewable Energy Systems: Benefits and Risks*, ed. David Pimentel (New York: Springer, 2008), 109–132.

David Ehrenfeld

1. Sir William Marris, trans., quoted in Michael Mott, "Strong Voices," *Poetry* 27, no. 3 (December 1975), 59–64.

2. Nassim Nicholas Taleb, *Fooled by Randomness: The Hidden Role of Chance in Life and in the Markets*, 2nd ed. (New York: Random House, 2005).

3. Charles Perrow, *Normal Accidents: Living With High-Risk Technologies*, 2nd ed. (Princeton, NJ: Princeton University Press, 1999).

4. Orrin H. Pilkey and Linda Pilkey-Jarvis, *Useless Arithmetic: Why Environmental Scientists Can't Predict the Future* (New York: Columbia University Press, 2007).

5. Paul Slovic, "Trust, Emotion, Sex, Politics, and Science: Surveying the Risk-Assessment Battlefield," *Risk Analysis* 19, no. 4 (1999).

R. James Woolsey

1. Sergey A. Zimov, Edward A. G. Schuur, and F. Stuart Chapin III, "Permafrost and the Global Carbon Budget," *Science* 312 (June 23, 2006), 1612–13.

2. Kenneth L. Denman, Guy Brasseur et al. "Couplings between Changes in the Climate System and Biogeochemistry," in *Climate Change 2007: The Physical Science Basis*, Fourth Assessment Report of the Intergovernmental Panel on Climate Change, eds. Susan Solomon et al. (Cambridge, UK: Cambridge University Press, 2007).

3. Piers Forster, Venkatachalam Ramaswamy et al., "Changes in Atmospheric Constituents and in Radiative Forcing," in *Climate Change 2007: The Physical Science Basis*, Intergovernmental Panel on Climate Change, eds. Susan Solomon et al. (Cambridge, UK: Cambridge University Press, 2007).

4. Fred Pearce, *With Speed and Violence: Why Scientists Fear Tipping Points in Climate Change* (Boston: Beacon Press, 2007), 77–85.

5. Ibid., 85 (citing L. Smith).

6. Jonathon T. Overpeck et al., "Paleoclimatic Evidence for Future Ice-Sheet Instability and Rapid Sea-Level Rise," *Science* 311 (2006), 1747–50.

7. Joseph Romm, *Hell and High Water: Global Warming: The Solution and the Politics—and What We Should Do* (New York: HarperCollins, 2007), 86.

8. Ibid.; Pearce, *With Speed and Violence*, 58.

9. Overpeck et al., "Paleoclimatic Evidence for Future Ice-Sheet Instability."

10. James Hansen et al., "Climate Change and Trace Gases," *Philosophical Transactions of the Royal Society* 365 (2007), 1925–54, http://pubs.giss.nasa.gov/docs/2007/2007_Hansen_etal_2.pdf.

11. J. E. Hansen, "Scientific Reticence and Sea Level Rise," *Environmental Research Letters* 2 (March 23, 2007), http://arxiv.org/ftp/physics/papers/0703/0703220.pdf.

12. Robert Zubrin, *Energy Victory: Winning the War on Terror by Breaking Free of Oil* (Amherst, NY: Prometheus Books, 2007), 109, quoting from Louis D. Johnston and Samuel H. Williamson, "The Annual Real and Nominal GDP for the United States, 1790–Present," *Economic History Services* (July 27, 2007); see also http://www.measuringworth.com/datasets/usgdp/result.php.

13. Ibid., 109.

14. Ibid., 109-10.

15. Florian Bressand et al., *Curbing Global Energy Demand Growth: The Energy Productivity Opportunity* (New York: McKinsey Global Institute, May 2007), 11.

16. Intergovernmental Panel on Climate Change, "Summary for Policymakers," in *Climate Change 2007: The Physical Science Basis.*

17. Pearce, *With Speed and Violence,* 59.

18. Romm, *Hell and High Water,* 86.

19. Ibid., quoting Peter Barrett.

20. Jay Gulledge, of the Pew Center on Global Climate Change, generated maps showing inundation patterns in May 2007. See appendix B in chapter 3 of Kurt M. Campbell, ed., *Climatic Cataclysm: The Foreign Policy and National Security Implications of Climate Change* (Washington, DC: Brookings Institute Press, 2008).

21. Roger Pielke et al., "Climate Change 2007: Lifting the Taboo on Adaptation," *Nature* 445 (2007), 597–98, citing Kelvin S. Rodolfo and Fernando P. Siringan, "Global Sea-level Rise Is Recognised, but Flooding from Anthropogenic Land Subsidence Is Ignored around Northern Manila Bay, Philippines," *Disasters* 30 (2006), 118–39.

22. J. L. González and T. E. Törnqvist, "Coastal Louisiana in Crisis: Subsidence or Sea Level Rise?" *Eos: Transactions of the American Geophysical Union* 87, no. 45 (November 7, 2006), 493-508.

23. Raymond S. Bradley et al., "Climate Change: Threats to Water Supplies in the Tropical Andes," *Science* 312 (2006), 1755-56.

24. National Research Council, *Making the Nation Safer: The Role of Science and Technology in Countering Terrorism* (Washington, DC: National Academy of Sciences, 2002), 182.

25. John S. Foster Jr. and others, *Report of the Commission to Assess the Threat to the United States from Electromagnetic Pulse (EMP) Attack* (Washington, DC: U.S. Electromagnetic Pulse Commission, 2004), 1-2.

PART FOUR FALSE SOLUTIONS

David Hughes

1. Vice-presidential debate between Sarah Palin and Joe Biden at Washington University, St. Louis, Missouri, October 2, transcript found at http://debate .wustl.edu/transcript.pdf.

2. R. A., "Michele Bachmann's Plan to Deliver Cheap Petrol," *The Economist*, August 18, 2011.

3. Daniel Yergin, "There Will be Oil," *Wall Street Journal*, September 17, 2011.

4. U.S. Geological Survey, "3 to 4.3 Billion Barrels of Technically Recoverable Oil Assessed in North Dakota and Montana's Bakken Formation—25 Times More than 1995 Estimate," *USGS Newsroom*, April 10, 2008, http://www .usgs.gov/newsroom/article.asp?ID=1911.

5. Eric Konigsberg, "Kuwait on the Prairie: Can North Dakota Solve the Energy Problem?" *The New Yorker*, April 25, 2011, http://www.newyorker .com/reporting/2011/04/25/110425fa_fact_konigsberg.

6. Walter Youngquist, "Shale Oil—The Elusive Energy," *Hubbert Center Newsletter* 98, no. 4 (October 1998), http://hubbert.mines.edu/news /Youngquist_98-4.pdf.

7. U.S. Energy Information Administration, *International Energy Outlook 2011*, "Projections of Liquid Fuels and Other Petroleum Products (2007-2035)," accessed December 9, 2011, at http://www.eia.gov/forecasts/ieo/excel /appe_tables.xls.

8. Ibid.

9. J. David Hughes, *Will Natural Gas Fuel America in the 21st Century?* (Santa Rosa, CA: Post Carbon Institute, June 2011), http://www.postcarbon.org /naturalgas.

10. U.S. Energy Information Administration, *International Energy Outlook 2011*, "Projections of Liquid Fuels."

11. Data from Arnulf Grübler, "Technology and Global Change: Data Appendix," accessed December 9, 2011, at http://www.iiasa.ac.at/~grubler /Data/TechnologyAndGlobalChange/; BP, *Statistical Review of World Energy 2011*, (June 2011), http://www.bp.com/statistical review; Energy Information Administration, *International Energy Outlook 2011*, "Projections of Liquid Fuels."

Richard Bell

1. Lucas W. Davis, "Prospects for Nuclear Power," National Bureau of Economic Research NBER Working Paper No. 17674 (December 2011), http://www.nber.org/papers/w17674.

2. Doug Koplow, *Nuclear Power: Still Not Viable without Subsidies* (Cambridge, MA: Union of Concerned Scientists, February 2011), http://www.ucsusa .org/assets/documents/nuclear_power/nuclear_subsidies_report.pdf.

Juan Pablo Orrego

1. Patrick McCully, *Silenced Rivers: The Ecology and Politics of Large Dams*, (London: Zed Books, 2001).

2. World Commission on Dams, *Dams and Development: A New Framework for Decision-Making* (London, Earthscan, 2000).

3. LeRoy Poff, Julian Olden, David Merritt, and David Pepin, "Homogenization of Regional River Dynamics by Dams and Global Biodiversity Implications," *Proceedings of the National Academy of Sciences* 104, no. 14 (April 3, 2007), 5732–5737.

4. World Commission on Dams, *Dams and Development*.

5. See Edward Goldsmith and Nicholas Hildyard, *The Social and Environmental Effects of Large Dams* (San Francisco: Sierra Club Books, 1984); Fred Pearce, *The Dammed: Rivers, Dams, and the Coming World Water Crisis* (London: Bodley Head, 1992); Patrick McCully, *Silenced Rivers: The Ecology and Politics of Large Dams*.

6. International Energy Agency, *Key World Energy Statistics* (Paris: IEA, 2010).

7. World Commission on Dams, *Dams and Development*.

8. Ibid.

Rachel Smolker

1. C. Nellemann et al., eds., *The Last Stand of the Orangutan—State of Emergency: Illegal Logging, Fire and Palm Oil in Indonesia's National Parks* (Arendal, Norway: United Nations Environment Programme and GRID-Arendal, 2007).

2. Jeremy Hance, "Rainforest Scientists Urge UN to Correct 'Serious Loophole' by Changing its Definition of 'Forest'," *Mongabay.com*, June 24, 2010, http://news.mongabay.com/2010/0624-hance_atbc_forests.html.

3. Edith M. Lederer, "Production of Biofuels 'is a Crime'," *The Independent*, October 27, 2007, http://www.independent.co.uk/environment /green-living/production-of-biofuels-is-a-crime-398066.html.

George Wuerthner, Oil Shale Development

1. United States Energy Information Administration, "Expectations for Oil Shale Production," 2009, http://www.eia.gov/oiaf/aeo/otheranalysis/ aeo_2009analysispapers/eosp.html.

2. World Energy Council, *2010 Survey of Energy Resources* (London, 2010); Ivan Sandrea and Rafael Sandrea, "Global Oil Reserves: Recovery Factors Leave Vast Target for EOR Technologies," *Oil & Gas Journal*, November 5, 2007, http://www.ogj.com/articles/print/volume-105/issue-41 /exploration-development/global-oil-reserves-1-recovery-factors -leave-vast-target-for-eor-technologies.html.

3. U.S. Geological Survey, "Assessment of In-Place Oil Shale Resources of the Green River Formation, Greater Green River Basin in Wyoming, Colorado, and Utah," Fact Sheet 2011-3063, June 2011.

4. U.S. Energy Information Administration, "Expectations for Oil Shale Production," 2009.

5. Harry R. Johnson et al., *Strategic Significance of America's Oil Shale Resource*, Office of Naval Petroleum and Oil Shale Reserves, U.S. Department of Energy (March 2004), http://fossil.energy.gov/programs/reserves/npr /publications/npr_strategic_significancev1.pdf.

George Wuerthner, Gas Hydrates

1. Woods Hole Science Center, "Gas Hydrate: What Is It?" U.S. Geological Survey, August 31, 2009, http://woodshole.er.usgs.gov/project-pages/ hydrates/what.html.

2. National Energy Technology Laboratory, "Fire in The Ice: Gas Hydrate Project Could Unlock Vast Energy Resource in Alaska," February 21, 2007, http://www.sciencedaily.com/releases/2007/02/070221180908.htm.

3. Marine and Coastal Geology Program, "Gas (Methane) Hydrates—A New Frontier," U.S. Geological Survey, 1992, http://marine.usgs.gov/fact-sheets /gas-hydrates/title.html.

4. National Energy Technology Laboratory, "Gas Hydrate Production Trial Using CO_2/CH_4 Exchange," May 29, 2012, http://www.netl.doe .gov/technologies/oil-gas/FutureSupply/MethaneHydrates/projects /DOEProjects/MH_06553HydrateProdTrial.html.

5. U.S. Geological Survey, *Assessment of Gas Hydrate Resources on the North Slope, Alaska, 2008* (Reston, VA: October 2008), http://pubs.usgs.gov /fs/2008/3073.

6. Woods Hole Science Center, "Gas Hydrate: Why Do We Study It?" U.S. Geological Survey, August 31, 2009, http://woodshole.er.usgs.gov /project-pages/hydrates/why.html.

7. Michael Benton, "Wipeout: When Life Nearly Died," *New Scientist* 2392 (April 26, 2003), http://www.newscientist.com/article/mg17823925 .000-wipeout-when-life-nearly-died.html.

8. Appy Sluijs et al., "Climatic Chain Reaction Caused Runaway Greenhouse Effect 55 Million Years Ago," December 21, 2007, http://www .sciencedaily.com/releases/2007/12/071221222544.htm.

9. Ray Boswell, testimony before the U.S. House of Representatives Committee on Natural Resources, Subcommittee on Energy and Mineral Resource, July 30, 2009, http://www.netl.doe.gov/newsroom/testimony /Testimony-NETL-Boswell-Final-7-27-09.pdf.

10. National Energy Technology Laboratory, "Gas Hydrate Production Trial Using CO_2/CH_4 Exchange."

11. Ray Boswell, testimony before the U.S. House of Representatives, July 30, 2009.

Brian L. Horejsi

1. Juliet Eilperin and Scott Higham, "How the Minerals Management Service's Partnership with Industry Led to Failure," *Washington Post*, August 24, 2010.

2. U.S. Government Accountability Office, *Energy Policy Act of 2005: Greater Clarity Needed to Address Concerns with Categorical Exclusions for Oil and Gas Development under Section 390 of the Act*, GAO-09-872 (Washington, DC, September 2009).

ETC Group

1. David Keith, "Engineering the Planet," in *Climate Change Science and Policy*, eds. Stephen Schneider et al. (Washington, DC: Island Press, 2009).

2. The Royal Society, *Geoengineering the Climate: Science, Governance and Uncertainty* (London, September 1, 2009), 62, http://royalsociety.org /policy/publications/2009/geoengineering-climate/.

3. See Institute of Mechanical Engineers, "Climate Change: Have We Lost the Battle?" November 2009, http://www.imeche.org/about/keythemes /environment/Climate+Change/MAG.

4. James R. Fleming, "The Climate Engineers," *Wilson Quarterly*, Spring 2007, http://www.wilsoncenter.org/index.cfm?fuseaction=wq.essay&essay_ id=231274 10; United Nations Framework Convention on Climate Change, *Clean Development Mechanism: 2008 in Brief* (November, 2008), 3, at http:// unfccc.int/resource/docs/publications/08_cdm_in_brief.pdf.

5. Ibid. The rest of this section relies heavily on Fleming's article.

6. Edward Teller, Lowell Wood, and Roderick Hyde, *Global Warming and Ice Ages: I. Prospects for Physics-Based Modulation of Global Change*, Lawrence Livermore National Laboratory (August 15, 1996), http://www.osti.gov/cgi-bin/rd_accomplishments/display_biblio. cgi?id=ACC0229&numPages=21.

7. P. J. Crutzen, "Geology of Mankind," *Nature* 415 (January 2002), 23.

8. Edward Teller, Roderick Hyde, and Lowell Wood, *Active Climate Stabilization: Practical Physics-Based Approaches to Prevention of Climate Change*, Lawrence Livermore National Laboratory (April 18, 2002), http://www.osti.gov/cgi-bin/rd_accomplishments/display_biblio. cgi?id=ACC0233&numPages=10.

9. P. J. Crutzen, "Albedo Enhancement by Stratospheric Sulfur Injections: A Contribution to Resolve a Policy Dilemma?" *Climatic Change* 77 (2006), 211-219, http://www.cogci.dk/news/Crutzen_albedo%20enhancement _sulfur%20injections.pdf.

10. Publication searches were conducted August 25, 2009. For scholarly articles: Google Scholar for the years 1994-2001 and 2002-present (search terms "geoengineering" and "climate" "change" in the following catego ries: Biology, Life Sciences, and Environmental Science; Chemistry and Materials Science; Engineering, Computer Science, and Mathematics; Physics, Astronomy, and Planetary Science; Social Sciences, Arts, and Humanities. For major media coverage: Lexis Nexis for the years 1994-2001 and 2002-present (search terms "geoengineering" "climate" "change") in newspapers stories (major world newspapers), weblogs and magazines.

11. William J. Broad, "How to Cool a Planet (Maybe)," *The New York Times*, June 27, 2006.

12. Seth Borenstein,"Global Warming is so Dire, the Obama Administration is Discussing Radical Technologies to Cool Earth's Air," *Associated Press*, April 9, 2009, http://abcnews.go.com/Technology/story?id=7313752&page=1#. UAS3444ftvR.

13. Steven Chu discussed geoengineering at the St James's Palace Nobel Laureate Symposium in London held on May 26-28, 2009.

14. Newt Gingrich, "Stop the Green Pig: Defeat the Boxer-Warner-Lieberman Green Pork Bill Capping American Jobs and Trading America's Future," June 3, 2008, archived at http://www.humanevents.com/2008/06/03/ stop-the-green-pig-defeat-the-boxerwarnerlieberman-green-pork-bill -capping-american-jobs-and-trading-americas-future/.

15. Alex Steffen, "Geoengineering and the New Climate Denialism," Worldchanging.com, April 29, 2009, http://www.worldchanging.com /archives/009784.html.

16. See, for example, Institute of Mechanical Engineers (UK), "Geoengineering: Giving us the Time to Act," August 2009, http:// www.imeche.org.

PART FIVE UNDER ATTACK

Erik Molvar

1. Hall Sawyer et al., *Sublette Mule Deer Study (Phase II): Long-Term Monitoring Plan to Assess the Potential Impacts of Energy Development on Mule Deer in the Pinedale Anticline Project Area*, 2005 Annual Report (Cheyenne, WY: Western Ecosystems Technology, Inc., 2005), http://www.west-inc.com /reports/PAPA_2005_report_med.pdf.

2. F. W. Lindzey, "Piney Front Elk Study," (presentation to the Governor's Planning Office, Wyoming, 2005).

3. Walter Merschat, declaration in exhibit to *Biodiversity Conservation Alliance v. Bennett*, Appeal from the Record of Decision, Atlantic Rim Natural Gas Development Project, Interior Board of Land Appeals IBLA 2007–210, U.S. Department of the Interior.

4. J. N. Dull, *Documentation and Appraisal of Known Gas Seeps within the Atlantic Rim Coal Bed Natural Gas Development Area, Carbon County, Wyoming*, U.S. Department of the Interior, Bureau of Land Management Rawlins Field Office, (February 6, 2007).

Winona LaDuke, with Martin Curry

1. Richard Girard, *Out on the Tar Sands Mainline: Mapping Enbridge's Web of Pipelines*, Polaris Institute (Ottowa: May, 2010), citing statistics obtained from Enbridge's "Environmental Health and Safety and Corporate Social Responsibility" reports, http://www.tarsandswatch.org/files/Updated %20Enbridge%20Profile.pdf.

2. Mark Brush, "Report: Enbridge Stopped and Restarted Pipeline During Oil Spill," *Michigan Public Media*, May 30, 2012, citing report released earlier in May 2012 by the U.S. National Transportation Safety Board, http:// www.michiganradio.org/post/report-enbridge-stopped-and-restarted -pipeline-during-oil-spill.

Michael Watts

1. Dauda S. Garuba, *Trans-Border Economic Crimes, Illegal Oil Bunkering and Economic Reforms in Nigeria*, Global Consortium on Security Transformation (Santiago, Chile: GCST, October, 2010), http://www.revenuewatch.org /news/economic-reforms-unfinished-nigeria-remains-exposed-oil -bunkering.

2. United Nations Development Programme/World Bank Energy Sector Management Assistance Programme, *Strategic Gas Plan for Nigeria* (February 2004), http://www.esmap.org/esmap/node/646.

Vandana Shiva

1. Charles Fishman, *The Wal-Mart Effect* (New York: Penguin, 2006), 102–103.

2. Ibid., 103.

3. New Economics Foundation, *Chinadependence: The Second UK Interdependence Report* (London: 2007), http://www.neweconomics.org/publications/chinadependence.

4. Christian Aid, *Coming Clean: Revealing the U.K.'s True Carbon Footprint* (London: February, 2007), 6, http://www.christianaid.org.uk/Images/coming-clean-uk-carbon-footprint.pdf.

5. ActionAid, *Meals per Gallon: The Impact of Industrial Biofuels on People and Global Hunger* (London: January 2012), www.actionaid.org.uk/doc_lib/meals_per_gallon_final.pdf.

6. Jayati Ghosh, "Frenzy in Food Markets," January 20, 2011, http://triplecrisis.com/frenzy-in-food-markets/.

7. Paul Wolfowitz, "The Clean Energy Challenge," (speech at Worldwatch Institute conference on Biofuels for Transportation, Washington, DC, June 7, 2006), http://go.worldbank.org/KOAHTX3L80.

8. Vandana Shiva, *Food vs. Fuel* (New Delhi: Navdanya, 2008).

9. Alex Salkever, "Global Biofuels Market to Hit $247 Billion by 2020," *Daily Finance*, July 24, 2009.

10. ActionAid, *Meals per Gallon*.

11. Food and Agriculture Organization, "Crop Prospects and Food Situation," 2009, http://www.fao.org/giews/english/cpfs/index.htm#top; Food and Agriculture Organization, "Food Outlook," 2009, http://www.fao.org/giews/english/fo/index.htm; ActionAid, *Meals per Gallon*.

12. Organization for Economic Cooperation and Development/ Food and Agriculture Organization, "Agricultural Outlook," 2010; Klaus Deininger and Derek Byerlee, *Rising Global Interest in Farmland: Can It Yield Sustainable and Equitable Benefits?* (Washington, DC: The World Bank, 2011), 15.

PART SIX DEPOWERING DESTRUCTION

Bill McKibben

1. Lauren Morello and ClimateWire, "Phytoplankton Population Drops 40 Percent Since 1950," *Scientific American*, July 29, 2010.

2. Thomas Friedman, "Want the Good News First?" *The New York Times*, July 27, 2010.

3. Matt Taibbi, "The Great American Bubble Machine," *Rolling Stone*, April 5, 2010.

4. The Breakthrough Institute, "Breakthrough Initiatives to Reinvent America," http://thebreakthrough.org/ideas.shtml, accessed December 14, 2011.

5. Ralph Nader, "A Tribute to David Brower, Champion of the Environment," *Boulder Daily Camera*, December 10, 2000, http://www.commondreams.org/views/121000-104.htm.

6. Mike Davis, "Welcome to the Next Epoch," TomDispatch.com, June 26, 2008, http://www.tomdispatch.com/post/174949/mike_davis_welcome_to_the_next_epoch.

7. Thomas Friedman, "We're Gonna Be Sorry," *The New York Times*, July 24, 2010.

8. Foreign Policy, "The FP Top 100 Global Thinkers: Number 78. Bill McKibben," December 2009, http://www.foreignpolicy.com/articles/2009/11/30/the_fp_top_100_global_thinkers?page=full.

9. James Hansen et al., "Target Atmospheric CO_2: Where Should Humanity Aim?" *Open Atmos. Sci. J.* 2 (2008), 217–231.

10. 350.org's Flickr page, http://www.flickr.com/photos/350org/sets/.

11. See http://energyactioncoalition.org/.

Sheila Bowers and Bill Powers

1. E-mail communication between Don Kondoleon, California Energy Commission Transmission Evaluation Program, and Bill Powers, Powers Engineering, January 30, 2008.

2. National Renewable Energy Lab, "PVWatts Performance Calculator for Grid-Connected PV Systems, Version 1," http://rredc.nrel.gov/solar/calculators/PVWATTS/version1, accessed September 21, 2010.

3. J. C. Molburg, J. A. Kavicky, and K. C. Picel, *The Design, Construction, and Operation of Long-Distance High-Voltage Electricity Transmission Technologies*, Argonne National Laboratory Technical Memorandum ANL/EVS/TM/08-4, November 2007, available at http://solareis.anl.gov/guide/transmission/index.cfm.

4. Joel Klein, Figure 8 in *Comparative Costs of California Central Station Electricity Generation*, California Energy Commission Report CEC-200-2009-07SF (January, 2010), 22; U.S. Department of Energy, Table 4-2 in *SunShot Vision Study*, Draft (May 28, 2010), 17, http://www1.eere.energy.gov/solar/sunshot/vision_study.html.

5. J. R. DeShazo and Ryan Matulka, *Bringing Solar Energy to Los Angeles: An Assessment of the Feasibility and Impacts of an In-Basin Solar Feed-in Tariff Program*, Los Angeles Business Council/ UCLA Luskin Center for Innovation (July 2010), 39, http://www.labusinesscouncil.org/online _documents/2010/Consolidated-Document-070810.pdf.

6. Amita Sharma, "$1 Billion May Be Needed for Lines Tied to Sunrise Powerlink," KBPS Radio, January 28, 2010, http://www.kpbs.org/ news/2010/jan/28/1-billion-may-be-needed-for-lines-tied-sunrise-po; Maureen Cavanaugh and Hank Crook, "Costs and Benefits of Sunrise Powerlink Vary by Community," KBPS Radio, May 5, 2010, http://www .kpbs.org/news/2010/may/05/costs-and-benefits-sunrise-powerlink-vary -communit; Katherine Tweed, "Tehachapi Renewable Transmission Project Completes Phase One," Greentech Media, May 5, 2010, http://www .greentechmedia.com/articles/read/phase-one-completed-in-tehachapi -renewable-transmission-project.

7. Richard Jasoni, Stanley Smith, and John Arnone III, "Net Ecosystem CO_2 Exchange in Mojave Desert Shrublands During the Eighth Year of Exposure to Elevated CO_2," *Global Change Biology 11* (2005), 749–756.

8. Robert Gleason, Murray Laubhan, and Ned Euliss, Jr., eds., *Ecosystem Services Derived from Wetland Conservation Practices in the United States Prairie Pothole Region with an Emphasis on the U.S. Department of Agriculture Conservation Reserve and Wetlands Reserve Programs*, U.S. Geological Service Professional Paper 1745, February 2008, http://pubs.usgs.gov/pp/1745/.

9. John McChesney, "Outlook Dims for Popular Energy-Efficiency Loans," National Public Radio, July 29, 2010, http://www.npr.org/templates/story /story.php?storyId=128700648.

Philip Cafaro

1. Intergovernmental Panel on Climate Change (IPCC), *Climate Change 2007: Mitigation -Technical Summary* (Cambridge, UK: Cambridge University Press, 2007), 107, http://www.ipcc.ch/publications_and_data/ar4/wg3/en /contents.html.

2. Ibid.

3. Ibid., 111.

4. Donald Brown et al., *White Paper on the Ethical Dimensions of Climate Change*, Rock Ethics Institute, Pennsylvania State University, 2007, http://rockethics.psu.edu/climate/.

5. Walter Reid et al., *The Millennium Ecosystem Assessment: Ecosystems and Human Well-being. Synthesis* (Washington, DC: Island Press, 2005).

6. Brian Czech, *Shoveling Fuel for a Runaway Train: Errant Economists, Shameful Spenders, and a Plan to Stop Them All* (Berkeley: University of California Press, 2002); James Gustave Speth, *The Bridge at the Edge of the World: Capitalism, the Environment, and Crossing from Crisis to Sustainability* (New Haven: Yale University Press, 2009).

7. IPCC, *Climate Change 2007: Mitigation*, Chapter 8, "Agriculture."

8. Ibid., Chapter 5, "Transport and Its Infrastructure;" see also IPCC, *Aviation and the Global Atmosphere* (1999).

9. Michael Huesemann, "Can Advances in Science and Technology Prevent Global Warming? A Critical Review of Limitations and Challenges," *Mitigation and Adaptation Strategies for Global Change,* 11 (2006), 566.

10. Bill McKibben, *Deep Economy: The Wealth of Communities and the Durable Future* (New York: Henry Holt, 2007).

11. N. Stern, *The Economics of Climate Change,* (Cambridge: Cambridge University Press, 2005).

12. Bill Ritter Jr., *Colorado Climate Action Plan: A Strategy to Address Global Warming* (Denver: Office of the Governor of Colorado, November 2007), 2, http://www.cdphe.state.co.us/ic/coloradoclimateactionplan.pdf.

13. Valuable contributions toward specifying the parameters of a sustainable economy include Herman Daly and John Cobb, *For the Common Good: Redirecting the Economy toward Community, the Environment, and a Sustainable Future* (Boston: Beacon Press, 1989); and Samuel Alexander, ed., *Voluntary Simplicity: The Poetic Alternative to Consumer Culture* (Whanganui, New Zealand: Stead & Daughters, 2009).

14. David I. Stern,"The Rise and Fall of the Environmental Kuznets Curve" *World Development* 32 (2004), 1419–1439.

15. Eileen Crist, "Beyond the Climate Crisis: A Critique of Climate Change Discourse" *Telos* 141 (2007), 29-55.

INDEX

EDITORS

TOM BUTLER is Editorial Projects Director of the Foundation for Deep Ecology. A Vermont-based activist and writer, he is the board president of the Northeast Wilderness Trust and the former longtime editor of the journal *Wild Earth*. His books include *Wildlands Philanthropy: The Great American Tradition* and *Plundering Appalachia: The Tragedy of Mountaintop Removal Coal Mining*.

DANIEL LERCH is Publications Director of Post Carbon Institute. He is the lead editor and manager of the Institute's major publications, including the *Community Resilience Guide* book series, a report series on shale gas, and the sixteen-author compilation *The Post Carbon Reader: Managing the 21st Century's Sustainability Crises*. He is the author of *Post Carbon Cities: Planning for Energy and Climate Uncertainty*.

GEORGE WUERTHNER is Ecological Projects Director of the Foundation for Deep Ecology. A photographer, author, and activist, he has published more than thirty books on America's wild places, including *Wildfire: A Century of Failed Forest Policy* and *Thrillcraft: The Environmental Consequences of Motorized Recreation*. He has served on the boards of several regional and national conservation organizations, and he lives in Helena, Montana.